Benjamin, Deleuze and the Baroque

Also Available from Bloomsbury

Media After Deleuze, David Savat and Tauel Harper
The Dialectics of Music, Joseph Weiss
Walter Benjamin, J. Roberts

Benjamin, Deleuze and the Baroque

The Early Modern Origins of Media Theory

Noa Levin

BLOOMSBURY ACADEMIC
LONDON • NEW YORK • OXFORD • NEW DELHI • SYDNEY

BLOOMSBURY ACADEMIC
Bloomsbury Publishing Plc
50 Bedford Square, London, WC1B 3DP, UK
1385 Broadway, New York, NY 10018, USA
29 Earlsfort Terrace, Dublin 2, Ireland

BLOOMSBURY, BLOOMSBURY ACADEMIC and the Diana logo are
trademarks of Bloomsbury Publishing Plc

First published in Great Britain 2025

Series design by Charlotte Daniels
Cover image: Woman looking into the hand mirror, stuccowork by Domenico
Beltramelli, Madonna del Popolo, Baroque, Cherasco, Province of Cuneo, Langhe,
Piedmont (© imageBROKER.com GmbH & Co. KG / Alamy Stock Photo)

A catalogue record for this book is available from the British Library.

A catalog record for this book is available from the Library of Congress.

ISBN: HB: 978-1-3504-1421-1
ePDF: 978-1-3504-1422-8
eBook: 978-1-3504-1423-5

Typeset by Integra Software Services Pvt. Ltd.

To find out more about our authors and books visit www.bloomsbury.com
and sign up for our newsletters.

'In the windswept towers of the Eiffel Tower, or better still, in the steel supports of a Pont Transbordeur, one meets with the fundamental aesthetic experience of present-day architecture: through the thin net of iron that hangs suspended in the air, things stream – ships, ocean, houses, masts, landscape, harbour. They lose their distinctive shape, swirl into one another as we climb onward, merge simultaneously' (Sigfrid Giedion, Bauen in Frankreich [Leipzig and Berlin] p.7). In the same way, the historian today has only to erect a slender but sturdy scaffolding – a philosophical structure – in order to draw the most vital aspects of the past into his net. But just as the magnificent vistas of the city provided by the new construction in iron (again, see Giedion, illustrations on pp. 61–3) for a long time were reserved exclusively for the workers and engineers, so too the philosopher who wishes here to garner fresh perspectives must be someone immune to vertigo – an independent, and if need be, solitary worker.

– Walter Benjamin, *The Arcades Project*

Contents

Introduction

> *a world, Solemque suum, sua sidera norat [that has its own sun, its own stars]*
>
> – Leibniz, *Theodicy*

> *We never know in advance how someone will learn: by means of what love someone will become good at Latin, what encounters make them a philosopher, or in what dictionaries they learn to think. The limits of the faculties are encased one in the other in the broken shape of that which bears and transmits difference.*
>
> – Deleuze, *Difference and Repetition*

i A strange encounter

In a rare mention of Walter Benjamin in one of his lectures at the Paris 8 Vincennes – Saint-Denis University, Gilles Deleuze dwells for a moment on Benjamin's Baroque theory. Deleuze highlights Benjamin's interpretation of allegories, used in German Baroque plays, as well as in paintings, sculptures and music, as a mode of expression that creates a rupture between the visual and verbal registers of the dramatic work; between the allegory and the allegorized.[1] A similarly disjunctive relation, Deleuze suggests, may be found between the excessively decorated external façades and the simple interiors, which nevertheless somehow fit together, of Baroque buildings.

Given in 1987, the final year before his retirement, to a packed roomful of students, the lecture belonged to one of the three seminars dedicated to Leibniz that Deleuze had given in the 1980s before drafting *The Fold, Leibniz and the Baroque*. In a recording of the lecture, one can almost hear the audience listening attentively. Deleuze is a natural storyteller, his voice animated, the fluctuating narration punctuated by dramatic pauses. It is no wonder his lectures were rumoured to be 'extraordinary'.[2] But when discussing Benjamin's *Origin of the German Trauerspiel* he stumbles for a moment, and then says, with some embarrassment, almost apologetically, that he has not managed to really enter the book. He quickly adds that it is a beautiful text and that others might have better luck doing so.

Although Deleuze may have had some misgivings about Benjamin's book, he returns to it in the final chapter of *The Fold*.[3] Benjamin's allegory, in fact, plays a crucial role in Deleuze's understanding of the Baroque. The relation, at once ruptured and continuous, encapsulated by Benjamin's definition of allegory is a precondition for the fold, the guiding concept through which Deleuze interprets Baroque science and culture. Deleuze quotes, in the opening of *The Fold*, Leibniz's elaboration of his principle of continuity through the image of folds in paper, such as those in a paper fan, that are separate from one another yet remain connected, forming one object together. The fan is opposed to a pile of grains of sand that are completely distinct from one another.[4] Deleuze and Benjamin both emphasize the discontinuities underpinning Leibniz's philosophical principle of continuity, the foundation for the infinitesimal calculus he developed independently from Isaac Newton. By doing so, they challenge the conventional ways of reading Leibniz's philosophy, and, arguably, diverge from the way in which Leibniz understood his own work.[5] The convergence between Benjamin and Deleuze, in their radical readings of Leibniz, is a rare one. Not only do they belong to different generations and intellectual traditions, they are also, as we shall see, two very dissimilar thinkers, and media philosophers. Both authors theorize and philosophize different forms of mass media, including film, photography and radio. Benjamin is one of the first to view forms of media such as the feuilleton, street advertisements and cartoons as deserving serious critical attention. In Deleuze's time, media theory was already an established discipline, but his two *Cinema* books broke new ground in presenting a rigorous philosophy of film unlike any at the time or since.

Deleuze was not the only one to struggle with Benjamin's 'Baroque book'. The obscure subject matter and unique form of this text, in which numerous lengthy quotations are inserted, were not well received by the gatekeepers of academia.[6] Benjamin had submitted it as his *Habilitation*, the formal qualification to become a university professor, only to be advised to withdraw the application in order to avoid official rejection. This event effectively changed the course of his career, from his envisioned academic one to that of a freelance literary critic. Unlike Deleuze, who found an academic home between the walls of the newly founded, experimental University Paris 8, Benjamin's academic destitution contributed to his increasingly precarious living conditions.

Written in 1924–5 and published in 1928, *Origin of the German Trauerspiel* is one of Benjamin's lesser-known works in the Anglo-Saxon context, although an excellent recent English translation might alter the situation.[7] The main theme of this work is the German Baroque *Trauerspiel*, a seventeenth-century dramatic form. With its often bombastic, over-the-top, blood-drenched plays, which typically take place in majestic courts and narrate the intrigues and politics of monarchical rule, it was considered an inferior, popular type of tragedy by the established literary critics of Benjamin's times. Against this approach, Benjamin argues that the Baroque *Trauerspiel* is an art form unto itself. Moreover, according to Benjamin, some of these lesser-quality works are as revealing, if not more so, for understanding Baroque drama than plays of critical acclaim.[8]

Similarly regarded as one of Deleuze's less accessible works, *The Fold, Leibniz and the Baroque* (1988) as its title indicates, paints a twofold philosophical portrait that is in some ways onefold: of the Baroque and of the scientist and philosopher Leibniz. The key concept of the fold is used to elaborate phenomena as diverse as Leibniz's calculus, physics and the concept of the soul on the one hand, and Baroque fashion and architecture on the other. While *Origin of the German Trauerspiel* was written rather early in Benjamin's career, *The Fold* belongs to the later stages of Deleuze's work. Yet in terms of subject matter, it seems to fit into a series of earlier philosophical monographs Deleuze had dedicated to figures like Friedrich Nietzsche, Immanuel Kant and Baruch Spinoza. *The Fold* differs from these monographs, however, in that in some ways it is also a collective portrait. Leibniz and the Baroque enter into dialogue with modern thinkers and artists whom Deleuze considers 'Leibnizian', ranging from Whitehead and Nietzsche, to Paul Klee, Paul Boulez and Jackson Pollock.

While in *The Fold* Deleuze distils his most detailed analysis of Leibniz's philosophy, he expressed a particular interest in Leibniz even before starting his university studies, and this interest persisted until his later years.[9] His first analyses of Leibniz's works were undertaken in *Expressionism in Philosophy: Spinoza*; *Difference and Repetition* (1968), which was submitted as his doctoral theses, and *Logic of Sense* (1969).[10] Deleuze's readings of Leibniz can be divided into two periods: an 'early period' in the works above, and a 'late period', twenty years later, in *The Fold* (1988) and briefly in *What Is Philosophy?* (1995). Following the prevalent distinction between Deleuze's earlier period ın which he wrote philosophical monographs and his later period in which he collaborated with Félix Guattari, the concerns that guide Deleuze's reading of Leibniz during the earlier and later phases are usually taken to be different. However, we will see this is not necessarily the case. During the interim period, moreover, Deleuze made use of the particularized conception of expression (without attributing it to Leibniz) in his collaborations with Guattari, *A Thousand Plateaus* (1980) and *Kafka, Towards a Minor Literature* (1975). He also taught three popular lecture-seminars on Leibniz at Paris 8.

Benjamin's interest in the Baroque is evident in his decision to dedicate his *Habilitation* to the quintessentially German *Trauerspiel*, or mourning play. He returns repeatedly to his theory of Baroque allegory, in the lengthy project he worked on intermittently during the last decade of his life, the *Arcades Project*. Leibniz's monads and *Monadology* are mentioned by Benjamin in texts ranging from his doctoral dissertation, 'The Concept of Criticism in German Romanticism' (written in 1919), to his final essay on the philosophy of history (1940).[11] Yet the only texts by Leibniz that Benjamin explicitly mentions, in the Foreword to *Origin of the German Trauerspiel*, are the *Monadology* and *Discourse on Metaphysics*, and he seems to confuse between the two.[12] There is also a reference to what Benjamin calls Leibniz's 'concept-language' in a bibliographical list composed as part of his studies on language.[13] Although based on partial knowledge, Benjamin's idiosyncratic and persistent engagement with Leibniz's conceptions, primarily the monad, is not negligible. Moreover, it is typical of the way in which Benjamin often read other authors and philosophers: appropriating specific elements from their philosophy and making them his own without necessarily crediting the appropriated source.[14]

Benjamin's reading of Leibniz was mediated to a certain extent by several neo-Kantian thinkers. As Schwebel notes, two sources appear to have informed Benjamin's early reading of Leibniz. Namely, the essay '*Leibniz' Weltanschauung als Ursprung seiner Gedankenwelt*' by Heinz Heimsoeth, mentioned in Benjamin's reading list, and Hermann Cohen's writing, specifically *Kant's Theory of Experience*.[15] While Cohen's reading was guided by an agenda of introducing Jewish sources into German philosophy, Heimsoeth, a student of Cohen's, linked Leibniz with Christian mystical traditions.[16]

Two further sources for Benjamin's understanding of Leibniz in the early years were his neo-Kantian professors Ernst Cassirer, whose course he attended between 1914 and 1915, and his professor Richard Herbertz.[17] Tagliacozzo and Ferrari suggest that Benjamin's preoccupation with the 'multiplicity and continuity of living forms', which links his work on Goethe with his understanding of Leibniz, was derived from Cassirer.[18] As McLaughlin mentions, Herbertz, who supervised Benjamin's doctoral dissertation, had written a study on Leibniz's theory of the unconscious.[19] Benjamin's late references to Leibniz's monad may have been mediated by the nineteenth-century philosopher Hermann Lotze, as will be discussed in Chapter 5.

Deleuze's references, on the other hand, demonstrate that he was versed in various editions of Leibniz's works as well as the literature on Leibniz, although it seems that he did not make much effort to stay up to date.[20] His reading of Leibniz was, however, notably mediated by several critics and writers, either his contemporaries of belonging to preceding generations, who were steeped in Leibniz's work. Michel Serres's *Le Système de Leibniz et ses modèles mathématiques* (1968) probably had the most decisive influence on *The Fold*. Deleuze quotes from Serres's work extensively; key aspects of Deleuze's interpretation of Leibniz, such as the analysis of Leibniz's perspectivism, are drawn from Serres. An earlier reading of Leibniz that was significant for Deleuze's understanding of Leibniz's conception of 'minute perceptions' was Salomon Maimon's *Essay on Transcendental Philosophy* (1790).[21]

Deleuze and Benjamin's receptions of Leibniz have served as the focal point of several recent studies.[22] Moreover, few studies examine the convergences between Deleuze and Benjamin's philosophies, and only one monograph interrogates Benjamin and Deleuze's work on Leibniz and the Baroque side by side.[23] This book views their respective appropriations of Leibniz's conceptions as grounds for a critical comparison between their philosophical works.

In recent years, perhaps because of a general renewed interest in Benjamin's sources, Benjamin's reading of Leibniz has received wider critical attention, yet most studies focus on the more explicit readings of Leibniz appearing in Benjamin's earlier work.[24] In the case of Deleuze, *The Fold* has received a fair amount of critical attention, serving as the object of multiple philosophical and architectural research projects. However, it is rare to find studies that consider how Deleuze's thought on Leibniz evolved and changed over time.[25] Contemporary readings of *Difference and Repetition* often ignore the fact that conceptions like 'intensity', 'virtual' and even 'expression' are informed by Leibniz's philosophy.[26] 'Intensity' in particular has been in vogue recently and yet it is not usually linked to Deleuze's reading of Leibniz.[27]

In terms of methodology, the book triangulates the three thinkers in order to offer new interpretations of each. While I argue for some affinities between the three, each philosopher belongs to a different historical period and cultural context and provides a different philosophical approach. It is through confrontations with these radically other ways of thinking that seldom-studied aspects of Benjamin, Deleuze and Leibniz's thought come to the fore.

I am interested in the ways in which Benjamin, Deleuze and Leibniz use a specific array of philosophical concepts – among them 'monad', 'expression', 'force' 'perspective' and 'image' – and in the function of these concepts in their philosophical projects more generally. I trace how the meaning of these concepts evolved and shifted through their usage and ask what these changing functions can tell us about each philosopher's philosophical approach. The works of all three thinkers will be interrogated in a trilateral reading that will serve as a living mirror revealing blind spots in the philosophies of each thinker.

As we will see, Benjamin and Deleuze's respective interpretations of Leibniz and the Baroque, and specifically Baroque concepts such as monad, expression and perspectivism, were crucial for the development of their media theories. In the Baroque we can therefore find, not a 'prehistory' but rather, in Benjamin's terms, a 'virtual history' of modern media, or even its 'deep history'.[28] In other words, the long past of mediatic expression as well as intimations of its possible futures may lie in the Baroque. These terms are especially suitable in light of Benjamin's view of Leibniz as a philosopher of immersion whose infinitesimal calculus penetrated deeply into reality.[29]

ii Intellectual lives

Born to a German-Jewish family in Berlin at the turn of the twentieth century, Benjamin developed his interests in literature, philosophy and politics from his early years. First, while attending a boarding school dominated by the influential pedagogue and youth movement reformist Gustav Wyneken, and then during his university years in Freiburg, Berlin and Bern, where he studied under the neo-Kantian philosopher Heinrich Rickert and the sociologist Georg Simmel.

From an early stage, as well, he wrote about various forms of cultural production: painting, literature, drama. His best-known writings on media, 'Little History of Photography' and 'The Work of Art in the Age of Its Technological Reproducibility', date from the later stages of his writing career, during the 1930s, but his interest in various forms of media and technologies predate these. *One-Way Street*, written in 1924–5, in part during the same years as *Origin of the German Trauerspiel*, explores a myriad of nascent modern media, from the typewriter to street advertising and photomontage. Although these two books, the only ones published in his lifetime, are different from one another in themes and style, both challenge common distinctions between 'high' and 'low' art, and show a deep interest in the overlooked, the marginal and the minute.

The First World War, and specifically the double suicide of his close friend Fritz Heinle and his lover Frederika Seligson as an act of protest against the war, profoundly affected Benjamin, driving him to develop one of his main themes in the 1930s, namely the ways in which the war brought about technological shifts which transformed everyday human experience. As he writes in 1933,

> Experience had fallen in value, amid a generation which from 1914 to 1918 had to experience some of the most monstrous events in the history of the world … a generation that had gone to school in horse-drawn street cars now stood in the open air, amid a landscape in which nothing was the same except the clouds and, at its centre, in a forcefield of destructive torrents and explosions, the tiny, fragile human body.[30]

After the war he sought an academic career, unsuccessfully, and was consequently forced into the precarious career of a literary and cultural critic. Benjamin spent the final decade of this life working on the monumental *Arcades Project*, better described as a 'work' than a 'project', as in its German title, *Passagenwerk.* A montage of quotations and fragments dedicated to the city of Paris, it remained incomplete at the time of his untimely death, leaving the question of its intended form open (and contested until today). Aside from the two books he published in his lifetime, in which he also conducted formal experimentation, most of Benjamin's literary output is written in the form of shorter texts: fragments, essays and reviews. Benjamin's itinerant existence and tragic ending lies in sharp contrast to Deleuze, who held a steady university position and authored more than twenty books in his lifetime.

Benjamin is sometimes described as belonging to the Frankfurt School. However, although in the last years of his life his main income was supplied by the Institute of Social Research, his work did not always suit its agenda. He shared the social critique of modernity, and, in particular, of capitalist society, with Critical Theorists such as Theodor W. Adorno, who was his close friend, Max Horkheimer, and other first-generation Frankfurt School researchers. While informed by Karl Marx and historical materialism, especially from the late twenties and on, Benjamin was not an orthodox Marxist, and critics continue to question the extent of his familiarity with Marx's works.

A long-lasting scholarly debate still exists between those who view Benjamin as, first and foremost, a historical materialist and those who think he was above all a theologian. To some extent the debate originates in Benjamin's correspondences. In their correspondence, Benjamin and Gershom Scholem held ongoing discussions of the changing focus of Benjamin's work over the years.[31] At times, Scholem confronted Benjamin with his shift away from a concern with Judaism and towards Communism.[32] Similarly, while Adorno's own work was marked by Benjamin's, he did not hold back his criticism of some elements in Benjamin's writing, as evident from their correspondence, in which he criticizes passages in Benjamin's *Arcades Project* as lacking in mediation and interpretation, or, in short, as not properly Marxist. 'Your work settled the crossroads of magic and positivism', as Adorno writes.[33] Yet this controversy falsely defines the 'materialist' and the 'theological' as oppositional poles, or forces, in Benjamin's writing, when, in fact, it is part of the particularity of his work that they complement each other.[34]

Some of Deleuze's contemporaries, like Jacques Derrida and Paul de Man, were perhaps informed by Benjamin to a larger extent than Deleuze. But this book is less concerned with the reception of Benjamin's work. Rather my interest lies in Benjamin and Deleuze's fascination with the Baroque and with Leibniz's philosophy. Unlike Benjamin, Deleuze, born in 1925, enjoyed a rather classical academic career, growing up in Paris in what he describes as a bourgeois family who were politically conservative, living through the pre-war crisis and studying history of philosophy at the Sorbonne in the post-war years.[35] His work is often, as, for example, in a conversation with Raymond Bellour and François Ewald from 1988, divided into three chronological periods. His early books were monographs dedicated to philosophers. Starting with his doctoral dissertation on Hume, published as *Empiricism and Subjectivity* (1953), this period then includes books devoted to Bergson, Spinoza, Kant and Nietzsche. His selection of philosophers in this period is often viewed as untraditional, or 'heretical', choosing to focus, for example, on empiricism when most of his colleagues were working on phenomenology.

In the second 'period', starting in 1968 and extending into the 1970s, according to this narrative, Deleuze developed his own philosophy in works such as *Difference and Repetition* and *The Logic of Sense* as well as works written with Félix Guattari, such as the two volumes of *Capitalism and Schizophrenia*. The third period, which relates to books written in the 1980s, was dedicated to art, and specifically to the image, with a book on Francis Bacon and the two *Cinema* books.

Deleuze contests, however, in the interview, this periodization and its implication that he worked on distinct projects, rather suggesting that the various tendencies of his work 'lead into one another and get mixed up'.[36] Thus, the monographs of Deleuze's first period are unorthodox in the sense that in as much as they focus on the philosophy of individual philosophers, they also create and put forward concepts and ideas that are distinctly Deleuze's own. In other words, they 'may best be conceived as a creative encounter between two philosophers'.[37]

The books on art written during the 'third period' are also philosophical treatises – one can argue that they elaborate Deleuze's philosophy to the same extent as the works belonging to the 'second period'. In the case of *The Fold* this is especially the case since it is, at the same time, a book dedicated to a philosopher, a book on (Baroque) art and culture and very much Deleuze's 'own philosophy'. The book thus embodies his conviction that philosophy, which operates through concepts, is inseparable from non-philosophy, specifically art, that operates with precepts and affects.[38]

Deleuze found an academic home, from 1969, at the experimental Paris 8 University, founded in the wake of the May 1968 protests, to which he was invited by Michel Foucault, who headed the philosophy department. There he enjoyed what he describes as ideal teaching conditions, in which he could delve into topics of his choice for several years, with a relatively fixed audience composed of philosophers and non-philosophers alike. 'A course is a kind of *Sprechgesang*', as he described it, 'closer to music than to theatre. Indeed there is nothing in principle to stop courses being a little like a rock concert'.[39]

'Deleuze was, finally, the most "antisecurity" philosopher one can imagine, the most anticonformist, the most corrosive, the most refractory to all attempts of destruction (of culture, of nature, of man) that have become the common lot of our age – an age whose onset he has foreseen', writes the historian Elisabeth Roudinesco, who took part in his seminars.[40] Despite, or perhaps on account of, his anticonformism and subversive way of philosophizing, he also belonged to the 'turbulent times', as she describes them, of post-1968 French thought. Like other thinkers of this period, Deleuze was preoccupied by questions of freedom on the one hand and of structure on the other, she writes.[41] This is especially apparent in his reading of Leibniz in *The Fold*, in which in some ways the tension between liberty and necessity also frames the entire book.

As an anticonformist philosopher Deleuze did not identify himself with either of the two significant movements of post-war French thought, structuralism and poststructuralism, although it can be argued that, in different periods, his work belonged to both and that he took part in the historical 'turn', as it is often described, against structuralism. Unlike the Frankfurt School, neither of the two movements had a defined agenda, nor were they affiliated with a specific institution, and their boundaries are still realigned and debated today. In recent years some have argued that the two movements are closer to one another than they have been considered in the past, with many previously considered 'poststructuralist' authors being still guided by 'structuralist' questions.[42]

Borrowing methods from linguistics, structuralism focused on the relationships between elements in a system, seeking to isolate and apply ahistorical, universal, models and structures to the social sciences and humanities. In his essay 'How Do We Recognize Structuralism?' (written in 1967 and published in 1972) Deleuze numbers several attributes of structuralism. It has often been noted that some of these, such as the shift 'from the subject to practice' and the 'differenciator, differentiation', correspond to his own work in *Difference and Repetition* and *Logic of Sense*.[43] Yet this text also contains a first critical reassessment of structuralism, and it was followed by its explicit rejection by Deleuze and Guattari in *Anti-Oedipus* and *A Thousand Plateaus*, which were marked by the key concepts of 'war machine' and 'rhizome'.

Deleuze's preoccupation with structuralism had not disappeared by the time he wrote *The Fold*, but it is not central to this work. Although he draws on Serres' structuralist reading of Leibniz, which examines Leibniz's philosophy as a system, understanding his metaphysics through his mathematics, Deleuze reads Leibniz through a multiplicity of Baroque cultural and mathematical features, challenging the systematicity of Leibniz's philosophy as a whole.[44]

iii Redefining the Baroque

As Deleuze mentions, many researchers have questioned the consistency of the notion of the Baroque, while some have downright denied it exists.[45] Nevertheless, its influence on European culture is unquestionable, and it continues to exert its fascination on scholars to this day, as evident by the many volumes, exhibitions and conferences dedicated to the Baroque every year. Every few decades, it seems, the Baroque enjoys a renaissance.[46]

The most common etymology of the Baroque links it to an irregularly shaped pearl, its beauty exceeding conventional norms. This etymology, whether it provides the real provenance of the term or not, corresponds to the conventional, art-historical way of understanding Baroque aesthetics and style as excessive, distortive or disjunctive. Benjamin and Deleuze's readings, even when drawing on Baroque art historians such as Heinrich Wölfflin and Erwin Panofsky, do not comply with traditional definitions. Both highlight the significance of minute details rather than bombastic grandness within Baroque thought and culture, for example. They also challenge the traditional distinction between Mannerism and Baroque. Baroque is often dated from the beginning of the seventeenth century and defined in contrast to Mannerism, a sixteenth-century style of art and architecture. The term originates in the Italian word '*maniera*', which means style, or manner, and was used in the seventeenth century to describe the artificial style of sixteenth-century artists.

In part, Deleuze and Benjamin challenge these distinctions because they view the Baroque as more than a specific historical and cultural period; more even, than a style that transcends periodization. For Benjamin and Deleuze, beyond indicating a historical period, the Baroque functions as a meta-idea or concept. For Benjamin, modernity originates in the Baroque, yet one must keep in mind his very specific definition of origin as 'not the coming-to-be of what has originated but rather what originates in the becoming and passing away'.[47] That is, the Baroque cannot be equated with 'early' modernity, or the genesis of modernity, but rather consists of an open concept that develops alongside the coming-into-being of modernity. In a similar vein, Benjamin likens the Baroque to a mirror image of the contemporary, be it the romantic, modern or any other period. As he writes: '... in such brokenness and inner strife, the present age mirrors certain aspects of the spiritual constitution of the Baroque, right down to the details of artistic practice'.[48] modernity and the Baroque mirror one another's collective existential crises as well as the artistic practices that developed in response to these crises.

Deleuze and Benjamin acknowledge the Baroque, as well, as a specific, revolutionary turning point in the history of philosophy and science that led up to the Enlightenment. An age marked by theological and political upheavals that followed the crumbling of the dominant theological structures and the ensuing European wars of religion, the Baroque is often described as suffering a collective crisis in faith that brought about an existential crisis. The period in discussion, also described as the 'Counter-Reformation' – the Catholic resurgence in the wake of the Protestant Reformation – extends between the mid-sixteenth century and the beginning of the eighteenth.

These profound transformations are the grounds for both authors' conceptualizations of the Baroque as a rupture in historical time.[49] In its mode as a crisis or rupture of history, the Baroque prefigures modernity, viewed at one and the same time as partaking in an age of crisis and as the crisis in itself. 'In antithesis to the historical ideal of the Restoration, the Baroque is faced with the idea of catastrophe', Benjamin writes.[50] He describes an age in which eschatological ideas are abandoned and the hope for salvation in the next world, dominant during the Renaissance, is replaced with the realization that there is no beyond.

Deleuze similarly describes the Baroque as a 'long moment of crisis, in which ordinary consolation no longer had much value'.[51] In his attempt to compensate for this collapse of the world as people knew it, and the enormous existential crisis that ensued, Deleuze writes, Leibniz had to rebuild the world. His reconstruction of the world pushed to its extreme the optimistic idea of our world as the best – most perfectly rich in detail and complex – of all possible worlds.

These conceptions of Baroque as temporal rupture and Leibniz as God's schizophrenic attorney, as Deleuze describes him, provoke the conventional narratives of Leibniz as the forefather of Enlightenment concepts of progress and teleological history.[52] Reading their Baroque philosophies in conjunction, we will see how Benjamin and Deleuze's respective theories of visual culture are intertwined with critiques of Enlightenment historiography and politics. For Deleuze and Benjamin, rather, both the seeds of Enlightenment ideologies and the possibilities for their critiques may be found in Leibniz. Rather than viewing Leibniz's philosophy as forming part of a genealogy of Enlightenment ideas of historical progress, they draw on his concepts in order to productively reconceptualize the Baroque.

In both Deleuze and Benjamin's cases, the engagement with the Baroque and with Leibniz was not unique in their intellectual milieus. Many readings of Leibniz and the Baroque in the twentieth century are genealogies of modernity, and Benjamin and Deleuze's works are both cases in point. *The Fold* followed several significant philosophical inquiries into Leibniz, from Louis Couturat and Bertrand Russell's influential works, to his friends and colleagues Yvon Belaval, Martial Guéroult and Michel Serres.[53] What set Deleuze's reading of Leibniz in *The Fold* apart from other such 'returns' is that it is done *through* the Baroque and the concept of the fold, as has been previously noted.[54] Deleuze viewed *The Fold* as a bringing together and merging the different aspects of his work, especially the philosophy of history and the philosophy of art.[55] Unlike his contemporaries and predecessors, who focused pretty much exclusively on Leibniz's philosophy, for Deleuze this philosophy is inseparable from the art, architecture, science and culture of the Baroque.

The plays written by Second Silesian School in the later seventeenth century that were the focus of Benjamin's dissertation were, similarly, not an obscure subject matter as it may seem today. On the contrary, Newman suggests that the corpus Benjamin studied 'belonged to a complex network of texts and ideological positions associated with a period very much at the centre of contemporary debates about nationalism and modernity'.[56] As she highlights, Benjamin aptly converses with the critical literature on a uniquely German genre, the *Trauerspiel*, and its role in the definition of German modernity.[57]

To a certain extent, as we shall seek to demonstrate in this book, just as Deleuze's reading of Leibniz's philosophy is distinct in that it is done through the Baroque, Benjamin's reading of the German *Trauerspiel* is unique in that it is performed through Leibniz's philosophy (although this is not explicitly stated nor systematically undertaken). Benjamin's reading also challenges other accounts of the period by suggesting that the *Trauerspiel* is not an inferior Greek tragedy, as many before him held, and by highlighting the allegory as a privileged mode of Baroque expression.

Benjamin and Deleuze share several art-historical sources on the Baroque. Among these, Wölfflin's influential *Renaissance und Barock* (1888) was significant to both. In what has since become a classic art historical text, Wölfflin defined a unique 'Baroque style' of sixteenth- and seventeenth-century architecture in Italy, distinguishing it sharply from the Renaissance.[58]

Deleuze draws, in *The Fold*, on Wölfflin's analysis of Baroque architecture in terms of material traits for his central image of the Baroque two-storied house.[59] Benjamin, who attended Wölfflin's university lectures, was critical of both his teaching skills and the formalistic aspects of his art theory.[60] Even so, Benjamin credits Wölfflin as 'the first to understand the Baroque in a positive light'.[61] Other classic texts on the Baroque that both refer to include Panofsky's analysis of Albrecht Dürer and Alois Riegel's *Late Roman Art Industry*.[62] While Deleuze and Benjamin are informed by these important texts, both contended, as mentioned earlier, that the art-historical method must be supplemented by a philosophical one and the Baroque must not be reduced to an artistic style.

Last but not least, both authors designate the Baroque optic as a pre-cinematic style. Their readings are attentive, as well, to the epistemological, theological and political stakes of the art and science produced in the period. Leibniz's monad, as 'living mirror of the universe', encapsulates a type of cinematic experience that Deleuze describes as the 'crystal-image'. As Deleuze notes, Leibniz's monad anticipates digital technologies, as well, and its logic of expression can be described as the logic of coding and information.[63] Leibniz's pyramid, in the fable recounted at the end of the *Theodicy*, functions as an original multiverse, containing a multiplicity of virtual worlds that are reminiscent of the multiple miniature worlds that surround us on the screens we carry with us every day.

Deleuze and Benjamin's readings of Leibniz's philosophy as an origin of modern media and cinema can direct us to tracing the Baroque at the origins of our postmodern, post-cinematic mediatic environment.[64] Thus the pre-cinematic can help us better understand our own post-cinematic age, the new media of which have been described as 'essentially digital, interactive, networked, ludic, miniaturized, mobile, social, processual, algorithmic, aggregative, environmental, or convergent, among other things'.[65] For both authors, the return to Leibniz is a return to a pre-Enlightenment mode of affect and sensation that results from a continuity between body and mind, subject and object, nature and history in an anti-dualism that Benjamin terms 'natural history'. For Deleuze, Leibniz's structure of expression, according to which 'the world is infolded in each soul, but differently, because each illuminates only one little aspect of the overall folding', anticipates the 'the move towards replacing the system of a window and a world outside with one of a computer screen in a closed room'.[66] That is, Leibniz's fold predicts the digital age, and his 'mannerist' perspectivism is extended into what Deleuze names 'monstrous' perspectivism, which is distinctive of modern literature and cinema, and has only been extended and multiplied in its postmodern instantiations.

iv Leibniz, paradigmatic Baroque philosopher

Benjamin and Deleuze's original interpretations of the Baroque hinge on their subversive readings of Leibniz. For all their differences, possibly the most important element that brings Benjamin and Deleuze's 'Baroque books' together is their shared conviction about the role of Leibniz not only as the most important Baroque scientist and mathematician but also as paradigmatic Baroque thinker and philosopher. In this respect, their approach is highly unlike the prevalent ridicule towards Leibniz's metaphysics and 'optimism' epitomized in Voltaire's *Candide*.[67]

Neither thinker purported to offer an 'objective interpretation' of Leibniz's work; instead, they put Leibniz's conceptions into the service of their own writing. Benjamin, especially, sometimes uses the term 'monad' almost as a trope and seems to have internalized Leibniz's terminology without having a profound knowledge of his philosophy. Deleuze, on the other hand, was well versed in Leibniz's canon. Still, in both cases, we are dealing with a seventeenth-century philosopher viewed from the perspective of the twentieth century; hence theirs are unavoidably anachronistic readings.

Benjamin and Deleuze's readings of Leibniz's philosophy are unique, for a number of further reasons. First, as mentioned, they spans the entirety of their writing career. Second, in both cases, their appropriations of Leibniz's conceptions are linked with their interrogation of 'meta' questions related to philosophical method, as well as to the relation between philosophy and other disciplines such as mathematics (and science more generally), art and history. The 'great philosopher of continuity' allows both authors to test out the relation between continuity and discontinuity or fragmentation (both ontological and historical), as well as those between immanence and transcendence, singularity and repetition, which constitute key issues for Deleuze and Benjamin. Third, as we will see in Chapter 1, Leibniz's conception of expression plays a central role in Benjamin and Deleuze's respective philosophical thought.

This conception forms part of both authors attempts to maintain a tense balance between continuity and discontinuity, singularity and totality, which is crucial for their understanding of the relations between experience and history (in Benjamin's case) or between experience and time (in Deleuze's). Deleuze and Benjamin turn to expression in their search for an alternative to the manner in which perception has been previously conceived of, namely as a structure of hierarchic representation. These hierarchic conceptions of experience and perception can be found, for example, in Aristotle's and Kant's philosophical systems.[68] Kant specifically is challenged by Deleuze and Benjamin for creating the chasm between the objective world outside and the way in which it appears to us.[69]

Unlike models of representation in which represented and representor are separate, their significance scaled by their distance from the truth, in Leibniz's model of expression there is a continuity, a gradation of force or scale of intensity, between the expresser, the expressed and expression itself. A dynamic structure of inclusion, or containment, Leibniz's conception of monadic expression designates a structure of relations in which the expressed is virtually included within the expresser. The expresser is singular and separate, yet at the same time continuous with the expressed, a 'virtual' constantly tending towards actualization.

Using Leibniz's terms, we may say that this structure solves the problem of the tension between the 'principle of continuity' and the 'principle of indiscernibles'. While the first principle determines that the world is composed of an infinity of continuous series, according to the second, every simple substance or monad must be different from every other.[70] That is, although all simple substances, or monads, are continuous with each other, they are also intrinsically distinct from one another. Leibniz reconciles the apparent contradiction by designating the monad as capable of expression. Every monad, he writes, expresses all other monads, as well as God and the universe. It is thereby continuous with the universe as a totality, yet also separate, enabling a 'multiplicity in the unity'.[71]

Aside from their function within Benjamin and Deleuze's understandings of history and ontology respectively, the reconfigurations of perception and experience that draw from Leibniz's concept of expression form part of each thinker's engagement with what Benjamin defines as 'the problem of presentation [*Darstellungsproblem*]', and what Deleuze describes as the 'search for new means of philosophical expression'.[72] Their persistent engagement with this issue is also manifest in Benjamin's conception of *Denkbild*, or thought/thinking-image, and Deleuze's *image de la pensée* – 'image of thought'. Both continually examine the relation between thought and its representation – specifically its representation in and over time.

By the refiguration of the conception of the image as based on the relation between virtual and actual, instead of that between real and semblance, Benjamin and Deleuze put forward what we might call a 'politics of the image'. Benjamin's reconfiguration of the relations between experience and history results in a political theory of historical experience that reformulates the relations between individual and collective history. Deleuze also purports to offer, in his transcendental empiricism, a politics, by replacing the hegemonic 'image of thought' with a non-representational concept of the image. The culmination of this recasting of the image lies in Deleuze's 'crystal-image' and Benjamin's 'dialectical image', both of which contain a double movement: between virtuality and actuality, and between abbreviation and totalization. These are conceptions of the image that are 'moving', or active, in the sense that they include a force field of irresolvable tensions. They are dialectical in a sense, yet they lack any kind of resolution.

In *What Is Philosophy?*, Deleuze names critique as one of philosophy's ills. The critic simply notes the vanishing or transformation of concepts, while the philosopher must actualize concepts, in effect creating them.[73] Expression does not form part of critique, but rather contributes to this creative conceptual process. The monad, accordingly, does not represent; it expresses, or, more precisely, it encodes and transcodes. Rather than a 'window on the world', the monad, like a painting by Rauschenberg, is an 'opaque table of information', or the darkened chamber of a *camera obscura*.[74] Deleuze uses these metaphors to illustrate that the monad's expression does not comprise a representation as in a landscape painting in which the painter attempts to 'copy' the landscape; rather, information or images are produced within the monad through a process of coding.

To Benjamin, on the contrary, expression consists of a kind of representation. In the 'Epistemo-Critical Foreword' to *Origin of the German Trauerspiel*, he appropriates the concept of the monad into his theory of interpretation and critique of works of art. As he writes, 'The idea is a monad – the pre-established representation [*Representätion*] of phenomena rests within it, as in their "objective interpretation".'[75] That is, the predetermined relation of expression between the idea and the phenomena it represents is equivalent to the latter's 'objective interpretation'. The 'objectivity' of interpretation depends, in other words, on the possibility of its totalization. Benjamin emphasises the distinctness of each idea-monad, its 'isolation' in relation to the totality. These characteristics are also used in Benjamin's later formulations of the monad in the *Arcades Project*, where the monad is described as interruptive of the continuum of history, and historian-philosophers are cast as monadically solitary.[76]

It may seem odd that Benjamin, whose philosophy of history highlights discontinuity, would draw upon Leibniz, who is renowned for his 'principle of continuity' according to which 'nature never makes leaps', and who views history as similarly continuous.[77] The category of origin [*Ursprung*] outlined in the Foreword relies rather on an internal crack or leap [*Sprung*], as Weber notes.[78] In the *Arcades Project* Benjamin highlights the discontinuity of history against the Historicists, writing for example that 'continuity in the presentation of history is unattainable'.[79] And yet he also describes therein the founding concept of historical materialism as 'actualization', which seems to imply that it relies on a continuous process of realization.

Furthermore, Leibniz's concept of the monad persists until Benjamin's last written text, 'On the Concept of History',[80] where he applies the metaphor of a seed (as we will see, a central metaphor for the relation of expression, which will also appear in the *Arcades Project*) to describe time and truth. 'The nourishing fruit of what is historically understood', he writes, 'contains time in its *interior* as a precious but tasteless seed'.[81] Time is described as an internal kernel within historical understanding, a virtual presence that may or may not sprout, actualizing its nourishing force.[82] Leibniz's concept of the monad, as well as the metaphor of the seed, is used specifically in the context of Benjamin's elaboration of historical experience. In Benjamin's view, I argue, the discontinuity of history presupposes a continuity of historical experience and understanding; in fact, the tension between the two is necessary. The monadic structure, implying a continuity between individual and total historical experience, preconditions the 'revolutionary chance in the fight for the oppressed past'.[83]

At the same time, Leibniz's apparently continuous system includes 'fits and starts', as Benjamin observes by highlighting the monad as discontinuous in its isolation from the infinity of other monads.[84] As we will see, these 'fractures' relate in large part to Leibniz's theories of perspective and the unconscious. The discontinuity at the heart of Leibniz's metaphysics is only one of the fractures in his apparently perfect system. Deleuze and Benjamin's readings of his philosophy accentuate its hallucinatory, dizzying and dark aspects, widening the gaps and unravelling the seams that hold its structure together.[85]

Unlike Benjamin or Deleuze, Leibniz self-described his philosophy as a system.[86] His ambitions were far reaching: his 'First Truths', for example, were intended to bring about a new world order.[87] Leibniz himself viewed his own philosophical project's systematicity, order and 'fitness' as one of its defining qualities. Yet unlike such philosophical systems as Kant or Hegel's, which are presented in thick volumes, much of Leibniz's 'system' is scattered in an array of short essays and over correspondences with important European thinkers of his time. His writings, produced in French, German and Latin, are dialogical and polyphonic.

His working patterns were similarly dispersed over different fields and projects. As his correspondent, author Bernard Le Bovier de Fontenelle wrote, trying to sum up Leibniz's prolific output in the eulogy dedicated to his friend, 'this almost perpetual jumble, which did not produce any confusion in his ideas, these abrupt and frequent transitions from one subject to another completely opposite subject, which did not trouble him, would trouble and confuse this history'.[88] His writing processes were apparently chaotic, generating cross-fertilization across multiple fields, even if the ideas that they contain adhere to the Cartesian yardstick of 'clarity and distinction'. In their readings of Leibniz's philosophy, Benjamin and Deleuze make manifest its inherent fragmentary, at times counter-systematic nature.

The unresolved tensions between continuity and discontinuity, unity and fragmentariness, disjunction and correspondence that are at the heart of Leibniz's monadology are crucial for Benjamin and Deleuze's theories of learning and understanding of time and space, and therefore also for their media theories. The book will unfold this argument in five chapters, travelling over three hundred years from Leibniz's 'living mirrors of the world' to the infinite mirrors of the internet in our post-digital 'age of anxiety'.[89]

Chapter 1, 'Of Monads and Mirrors: Leibniz's Monad in Deleuze and Benjamin', discusses Deleuze and Benjamin's appropriations of Leibniz's metaphysical concepts, starting with the monad. The chapter first briefly introduces the main tenets of Leibniz's philosophy and their relation to his other scientific and mathematical endeavours. Leibniz famously describes, in his *Theodicy*, two labyrinths, or profound questions in which reason tends to get lost: human freedom and continuity. While the first pertains to the tension between human liberty and a predetermination deriving from a theological system controlled by an omnipotent God, what today we would define as ethical and theological questions, the second 'labyrinth' contains questions, upon which Leibniz elaborated in his physical and mathematical research, concerning divisibility, motion and continuity of substances. In other words, the image of this second labyrinth encapsulates Leibniz's struggles with the definition of simple substances that in his later writing Leibniz calls 'monad'.

The chapter elaborates on the monad, defined by Leibniz as a 'living mirror of the universe', and his concept of expression; the structure that guides the relation between monad, world and God. After highlighting the significance of these concepts for Leibniz, the chapter will discuss their appropriations by Benjamin and Deleuze. Specifically, we will see how Leibniz's monad and expression inform the relation between experience and knowledge within Benjamin and Deleuze's critical epistemologies. By using Leibniz's conception of expression, Benjamin and Deleuze offer, in their earlier works – namely Benjamin's *Origin of the German Trauerspiel* and Deleuze's *Difference and Repetition* – radical critiques of traditional epistemological theories of experience and learning, specifically those of Descartes and Kant.

Chapter 2, 'Infinite Tasks of Learning: The Baroque-Inspired Critical Epistemologies of Benjamin and Deleuze', continues elaborating on Benjamin and Deleuze's theories of learning, as a foundation for their theories on history, politics and media that will be unfolded in the following chapters. The chapter suggests that Benjamin and Deleuze's reconfigurations of the relations between knowledge and experience, described in Chapter 1, serve as grounds for alternatives to key modern epistemological theories. Challenging Kant and Descartes' epistemologies, Benjamin and Deleuze offer theories of learning, viewed as a dynamic and infinite process, rather than as the acquisition or possession of knowledge, informed by Leibniz's conception of 'infinite analysis', and the post-Kantian conception of 'infinite task'. Deleuze and Benjamin follow attempts made by philosophers such as Hermann Cohen, Ernst Cassirer and Salomon Maimon, who return to Leibniz in order to rectify Kant's transcendental system. Yet while Benjamin's concept of expression is based on an interpretation of Leibniz's monad as a limitation of the infinite, Deleuze calls for an unlimited expression.

Chapter 3, 'Forces of Eternal Return: Baroque Forces in Benjamin's Theory of Historical Experience', considers the way in which Leibniz's conception of force, specifically 'living force', and the internal dynamic within the monad between virtuality and actualization figure within Benjamin's theory of historical experience. Leibniz's conception of virtuality is appropriated against concepts of progress developed by Kant and the Historicists, from which nationalist, colonialist and capitalist ideologies derived. By describing a force that, rather than striving for perfection or completion as in Leibniz and Hegel, is directed towards destruction, Benjamin critiques the violence that is necessary for a continuous and teleological history. Force forms part of the reconfiguration of the relations between experience and knowledge, discussed in Chapter 1, which extends into Benjamin's later theory of historical knowledge and experience. We will also see how Benjamin's concept of 'virtual history' relates to Leibniz's concept of virtuality, and how questions of virtual and multiple worlds inform questions of singularity and repetition, for Benjamin as well as Deleuze. The final section of the chapter will discuss the concept of eternal return, which plays an important part in Benjamin and Deleuze's understating of historical time that is informed, as well, by the emergence of modern concepts of history and historiography during the Baroque.

Chapter 4, 'It's all about Perspective: The Body Politics of the Baroque Image', interrogates how the revolutionization of visuality and perspective that took place during the Baroque informed Benjamin and Deleuze. By so doing, it lays the ground for the final chapter in which the Baroque is linked to twentieth-century media. The chapter begins by outlining key developments in geometry, optics and perspective by mathematicians such as Blaise Pascal and Girard Desargues, who contributed to the establishment of a new perspectival paradigm during the Baroque. It then shows the continuities between Leibniz's metaphysics and his geometric and mathematical research. This is most apparent in Leibniz's perspectivism, within which vision, perception, and body are linked through a theory of 'point of view' or perspective. Following Leibniz, Deleuze and Benjamin understand perspective and image as embodied, and vice versa: the body is constituted by its own perspective, which differentiates it from all other bodies. Both Deleuze and Benjamin pay special attention to Leibniz's emphasis on the 'in between', hallucinatory or intoxicated states of perception, within his theory of 'minute perceptions', understanding perception as constitutive of the body through differentiation. They appropriate Leibniz's perspectivism against hierarchical visual representation, as part of their respective body-political theories.

Chapter 5, 'From the Crystal Palace to Cinematic Crystals: Baroque Optic as Pre-cinematic Form', focuses on the different ways in which Baroque perspectivism is a precursor of cinematic form, according to Benjamin and Deleuze, and on the inspiration they find in the Baroque for their modern cinematic and photographic theories. The foundational role of Leibniz's perspectivism for modern cinema and literature is embodied in the myth of the crystal pyramid that seals the *Theodicy* and attempts to justify the mutual existence of divine justice and human freedom. The myth that suggests the co-existence of many virtual worlds informs Benjamin and Deleuze's understanding of virtuality as well as their key image-concepts, namely the 'dialectical image' and 'crystal-image'. While these concepts are similar in that they bring together image and ruptured temporality, they also render the divergences between Deleuze and Benjamin's thought most evident. While for Benjamin the dialectical image may offer historical truth and a path to revolution, Deleuze takes Leibniz's perspectivism a further step towards chaos and suggests that the crystal-image is essentially a falsifying mechanism. Here we return to arguments made in Chapter 3 that investigated Benjamin and Deleuze's shared critiques of Baroque-originated historicist historiographies. Despite these shared critiques, it will be shown that the ways in which they understand the interrelation between media and temporality remain irreconcilable.

The Conclusion summarizes the book's main arguments, clarifies their stakes and highlights their implications on contemporary theoretical debates. It links between Leibniz's perspectivism, its reception by Deleuze and Benjamin, and theories on perspective by contemporary authors such as Hito Steyerl and Eyal Weizman. Thereby, the present and possible future transformations of perspectivism in a post-photographic, post-cinematic world will be elaborated. The conclusion highlights the contributions made in the book to the reinterpretation of the philosophy of Leibniz and Baroque culture and science, and the re-evaluation of their roles in the history of philosophy and within media studies. Finally, it will emphasise the relevance of the Baroque, as understood by Deleuze and Benjamin, to our own age of multiple crises, revisiting Deleuze's definition of his times as 'neo-baroque'.

1

Of monads and mirrors: Leibniz's monad in Deleuze and Benjamin

Introduction: Colliding monadologies

During his first seminar dedicated to Leibniz, taught at Paris 8 University Vincennes – Saint-Denis, Deleuze singled Leibniz out as a philosopher whose penchant for infallible systems did not prevent him from creating strange concepts: 'very oddly in this taste for order and to establish this order, he yields to the most insane creation of concepts that we have ever witnessed in philosophy. Dishevelled concepts, the most exuberant concepts, the most disordered, the most complex in order to justify what is.'[1] Deleuze implies, provocatively, that the concepts Leibniz invented unsettle his philosophical system's apparent order from within.

This chapter begins, then, by exploring some of these concepts, such as the monad, continuity and pre-established harmony. Of course, there is a wealth of literature on each of these, and we can only introduce them here briefly, in a way that relates to Benjamin and Deleuze's readings. We will then trace the appropriations of these concepts by Benjamin and Deleuze in their earlier readings of Leibniz, and how they inform the relation between experience and knowledge within their critical epistemologies. Benjamin and Deleuze throw into question our most fundamental conceptions about the nature of knowledge and learning. Based on these critiques, we will see in Chapter 2, they put forward theories of learning that are informed by Leibniz's conceptions of 'expression' and 'infinite analysis'. By appropriating these concepts, Benjamin and Deleuze offer radical critiques of traditional epistemology and propose theories of learning and experience that are not subject-based.

Leibniz's concepts of 'expression' and 'infinite analysis' are grounded in his principle of continuity, which in turn is derived from his infinitesimal calculus.

The latter had inspired the neo-Kantian notion of 'infinite task'. Deleuze and Benjamin were both informed by this conceptual trajectory that emphasizes learning as a continual, open-ended process. The current chapter begins, then, by examining Leibniz's inquiry into substance that is undertaken in the context of Leibniz's two 'labyrinths', or key philosophical problems, that of human freedom and that of continuity. We then focus on the 'structure of expression' that determines and is determined by his novel understanding of the nature of substance.

Leibniz's concepts are interpreted by Deleuze and Benjamin in the context of their critiques of epistemology. Deleuze's first engagement with this concept appears in *Expressionism in Philosophy: Spinoza*, in which Spinoza and Leibniz's conceptions of expression are pitted against one another. In this work, Deleuze describes expression as an anti-Cartesian, immanent conception, identifying its rebellious nature within the history of philosophy. Benjamin discusses Leibniz directly only in the Foreword to *Origin of the German Trauerspiel*, but this chapter and the next will ground this discussion in Benjamin's earlier critiques of Kant's epistemology as well.

1 Leibniz's two labyrinths

i Freedom

Leibniz famously opens the *Theodicy* by describing two labyrinths in which

> our reason very often goes astray: one concerns the great question of the Free and the Necessary, above all in the production and the origin of Evil; the other consists in the discussion of continuity and of the indivisibles which appear to be the elements thereof, and where the consideration of the infinite must enter in. The first perplexes almost all of the human race, the other exercises philosophers only.[2]

The *Theodicy* is concerned with the first of these 'labyrinths'. As befitting a work written by a jurist, Leibniz's *Theodicy* defends the existence of a perfect, omnipotent and just God in a world in which evil and human suffering exist. The problem Leibniz struggles with is the compatibility of our God-created world as 'the best of all possible worlds' on the one hand, and the freedom of humans to sin on the other. In other words, how could a perfect God create what would seem to be a non-perfect world, in which evil exists? Leibniz's solution to this problem includes viewing evil as the limit of being and perfection.[3]

Leibniz's *Theodicy* has long been considered an inferior text of Leibniz's oeuvre. In the twentieth century especially, following the influential interpretations of Couturat and Russell, which prioritized Leibniz's logics above all other domains, Leibniz's theology has been relegated to a secondary place, despite the *Theodicy* being Leibniz's main philosophical work published during his lifetime.[4] However, as Rutherford argues, Leibniz's theological thinking and his metaphysics cannot be understood independently of each other. Deleuze seemed especially aware of this profound transdisciplinarity of Leibniz's system, in which the principles of logic, for example, are determined by metaphysical principles and vice versa.

What is sometimes described as Leibniz's 'optimism', or the idea that God chose our world as the best of all possible ones, is formulated in a theological context. According to this theory, 'this supreme wisdom, united to a goodness that is no less infinite, cannot but have chosen the best'.[5] In other words, as an omnipotent, omnibenevolent and omniscient being, God could only have created a perfect world.[6] It may seem that if He can only choose our world as the best of all possible worlds, His choice cannot be a free one. Yet Leibniz insists throughout the *Theodicy* that God chose freely, resolving the difficulty by redefining the concepts of freedom and necessity. Leibniz distinguishes between 'absolute necessity' and 'moral necessity', the latter which guides God's actions while not impeding His freedom. Furthermore, he argues, God's free choice is determined by knowledge of reason: 'The decrees of God are always free, even though God is always inclined to them by reasons which lie in the intention towards the good; to be necessitated morally by wisdom, to be bound by the consideration of the good, is to be free'.[7] This formulation enables Leibniz to insist on God's spontaneity as well as His possibility of selecting the best world among possible ones, which is the most harmonious and fitting of worlds.

As to the freedom of humans, Leibniz argues that it is assured by his distinction between truths that are 'primary' and 'necessary' and those that are 'contingent' or 'free'. 'Primary truths' are also known as 'identities', Leibniz writes, such as 'A is A' or 'A is not not-A'.[8] 'Derivative truths' are divided, in turn, into two groups. The first type of derivative truths, which Leibniz calls 'necessary truths', can be reduced into primary truths. The second type, 'contingent truths', cannot be reduced to primary truths; they are 'endowed with an infinite series of reasons, fully known to God alone', as Leibniz writes.[9] That is, 'contingent truths' cannot be demonstrated in a finite series of steps.

'Contingent truths' result from the limitation of human knowledge in comparison to God's perfect and infinite knowledge of the world. The idea that some truths require an infinite number of steps in order to be demonstrated is also known as Leibniz's principle of 'infinite analysis', which will be the focus of the next chapter. In an argument that may sound strange to modern eyes, Leibniz is in fact suggesting that humans are free because they are not in possession of God's infinite knowledge. If they were in the complete know, their actions would be predetermined. Leibniz describes, in his concise early exposition of his principles, known as 'First Truths', the 'complete individual' as containing all its predicates, past, present and future, yet doing so contingently, with the result that its actions are free.[10] Moreover, not only does Leibniz suggest that the concept of the individual includes all its present predicates but he also asserts that it includes its past and future predicates. Deleuze describes this as an 'identity of the predicate and the event'.[11] This means that for Leibniz, just as the individual contains its predicates, so humans 'contain' their future, as a virtual possibility.

ii Continuity

The second labyrinth, that of continuity, is pronounced as the most significant and most perplexing of philosophical problems. Continuity concerns, for Leibniz, both the metaphysical, ideal level and the physical levels of reality. It is the principle of organization of physical and metaphysical elements, according to which *Natura non facit saltum* (nature makes no leaps), or in other words, all things in nature are distinct from one another yet form a unified whole together.

As Leibniz mentions in the quote above, the question of infinity forms part of the consideration of the mathematical and physical problems of continuity, which enquire into the relation between indivisible elements. Since continuity concerns the organization of matter, Leibniz's conception of substance both determines and is determined by his principle of continuity and mathematical understanding of the infinite. The problem of continuity, according to Leibniz, thus concerns not only continuity of space, time and motion but also the questions of whether matter is divisible into first elements, and their motion.[12]

Leibniz exchanged Descartes' *rule*-based 'method', for a set of *principles* through which he constructs his system.[13] Instead of externally imposed, mathematical rules, Leibniz invented several basic principles that derived immanently from his philosophical, theological and mathematical beliefs, and which served him in his diverse research in various disciplines. These principles were described as the 'principle of sufficient reason', according to which everything has a reason or cause (a form of which, we shall see, was adopted by Deleuze); the 'principle of identity' and its reverse form, 'the principle of indiscernibles', according to which no two elements of nature may share all their predicates; and the 'principle of continuity', which was especially significant for Deleuze and Benjamin. Both philosophers sought in Leibniz, as mentioned in the introduction, a model or structure that allows both rupture and continuity. Benjamin defined this structure through the concept of allegory, Deleuze through that of the fold.

The principle of continuity pertains, first, to spatiotemporal continuity, determining that any change in time or space is always continuous, that is, passes through an infinity of intermediate degrees. It also argues for the continuity of existents, describing the succession of all creatures forming an uninterrupted chain. Finally, it is a law of continuity of reason, serving Leibniz as a *methodology* for reasoning, through which it is possible, when examining concepts that are continuous with one another, to determine that the laws applying to one must apply to the other. According to an account of the principle of continuity to which Leibniz returned throughout his writing,

> when the differences between two instances in a given series … can be diminished until it becomes smaller than any given quantity whatever, the corresponding differences in what is sought or in their results must of necessity also be diminished or become less than any given quantity whatever.[14]

In a series in which the difference between two instances 'approach each other continuously', that is, between variables in a continuous function, the corresponding difference in the result retains the same properties. The principle of continuity forms part of Leibniz's theory of knowledge as well as what he refers to as a 'psychology', the study of the mind.[15] These domains are construed through his conceptions of the individual substance, or monad.

2 The structure of expression

i Monad

The term 'monad' is a name given by Leibniz in his later work for what he earlier referred to as an 'individual' or 'singular' substance. 'Substance' is the term used since ancient Greek philosophy for the foundational elements or entities of the universe. As Leibniz writes in the *Monadology*, 'the monad, of which we will here speak, is nothing but a simple substance, which enters into compounds. By "simple" is meant without parts.'[16] The monad is an indivisible unit of which 'composites', or bodies, are composed. Leibniz is perhaps best known for his 'monadology', or theory of monads, even though his emphasis on the individual and coining of the term only came in the later stages of his thought. *Discourse on Metaphysics* (1686) is perceived as the first work in which Leibniz presents his mature conception of individual substance.[17] Yet it is only in the *Monadology* (1714), written just two years before his death, that he unfolds a complete metaphysics of the substance.

According to this elaborate metaphysical system, all created things, including humans, animals and plants, have souls, which he also refers to as the 'dominant monads' of the organism. However, only the souls of 'rational animals' are called 'minds'.[18] The bodies of all creatures are composed of an aggregate of inferior, non-rational monads. There is also a hierarchy of mind-monads with those of plants, or 'bare monads' at the bottom and those of God, humans, and angels at the top.

Leibniz's conception of substance is constructed through the problematic of continuity. For Leibniz, the individual substance is defined by its forming part of a continuum in which everything is in constant flux, and at the same time, by its indivisibility and its own relentless striving to actualize. On the one hand, Leibniz believed that all elements of nature form one unified continuum in which it is not possible to discern one element from the next.[19] On the other hand, he describes this continuum as composed of indivisible elements. Since individual substances are non-divisible, they are qualitative, which means they change internally and qualitatively.

Leibniz's understanding of substance is shaped against that of Descartes and his followers, who viewed matter as nothing other than extension, arguing that all changes in matter are reducible to motion.[20] Famously, Descartes believed that there are only two kinds, or categories of substance, or entities: matter, of which the principle quality is extension in space, and mind, of which the essential quality is thought. In other words, Cartesians are dualists and mechanists; that is, they offer mechanic explanations to the actions of both mind and matter. By contrast, Leibniz viewed matter as involving action.[21] Against Descartes' mechanistic view of substance, for Leibniz, an essential quality of any substance is its being endowed with an internal force that drives it to move and change.[22] These aspects of Leibniz's philosophy certainly attracted both Deleuze and Benjamin.

The source of these oppositional views of matter and substance lies in Leibniz and Descartes' dynamics. Indeed, Leibniz himself justifies the nature of substance, specifically its self-action, through recourse to his principle of conservation of force.[23] Against the Cartesian view of motion as external to bodies, Leibniz developed his principle of conservation of force, expressed by the formula *mv2*, according to which motion is understood as a change of position, and rest, or stillness, as an infinitely small quantity of motion.[24] Accordingly, force that brings about motion is an intrinsic quality of the substance, always present in the monad in virtual form and tending to actualization. The qualitative state of the substance is thus defined intensively by the ever-changing degree of its internal force. The individual substance is defined as intensive, since it is endowed with an active force of varying degree, motivated by a *conatus* or desire to actualize. Aside from its qualitative and intensive nature, a third key characterization of the monad defines it as expressive.[25]

The simple substance is defined by its capacity to express. In other words, it consists of an expressive 'centre'.[26] It is 'like a small world that expresses the large world', as Leibniz writes in *Discourse on Metaphysics*. The relation between the monad, which comprises a 'world' in itself, and the universe is that of expression (Leibniz plays on the similarity between *monde* and *monad*).

ii Expression

Leibniz never explains precisely what he means by expression and the term, as discussed, has elicited no shortage of controversy. That said, the significance of the term within his metaphysics and theory of knowledge is well established.[27] Expression is the linchpin of Leibniz's solution to the philosophical problems of continuity. It is through its capacity to express that the individual substance is continuous with the infinite continuum of elements that forms the universe, while also remaining distinct from this continuum in its indivisibility. As Leibniz describes in his *Monadology*, every monad, or simple substance, expresses the infinite totality of monads, representing this infinity from a different point of view. Within every single monad, there is constant expression of the totality of monads – expression is thus temporal and perpetual.[28]

Expression is also a meeting point of Leibniz's psychology, metaphysics and mathematics. As mentioned, human minds are viewed as monads, and are therefore expressive in a particular way that will be described below, in the section on perception and consciousness.

Leibniz often illustrates expression by way of geometrical examples:

> one thing expresses another (in my sense) when there is a constant and regular [*réglé*] relation between what can be said of the one and what can be said of the other. It is thus that a projection in perspective expresses the original figure.[29]

This means that, in expression, the constant element in both the expresser and the expressed is a law guiding the make-up of its inner relations, or proportions.

Leibniz describes the relations of expression as a congruence or proportional similarity between the expresser and the expressed. He often uses as an example the way in which a circle is expressed by its algebraic equation.[30] Yet the extent of this similarity remains a subject of contention.[31] His paradigmatic example for expression is the relation between a geometrical figure and its projection.[32] This example implies that expression is a relation of representation, in which the part represents the whole, and there is an isomorphic relation between the expresser and the expressed. Other, more abstract, examples for expression include the expression of the machine by the model of the machine, and the expression of thoughts and truths by speech.[33]

Leibniz often describes expression as a kind of mirroring, as in 'Principles of Nature and Grace', where he writes that 'each monad is a living mirror, or a mirror endowed with inner activity, representative of the universe, according to its point of view'.[34] The monad is described as a mirror that, led by its inner active force, reflects the universe in all its changes. One of the defining qualities of the monad, then, is its capacity to express or mirror the entire world, containing its minute mirror image, or double. The stakes of the concept of 'living mirrors' for Leibniz, Deleuze and Benjamin will be discussed further in Chapter 4.

iii Perception

'Perception', which is the term Leibniz uses when describing the psychological aspects of the relation of 'expression', plays a key role in his understanding of knowledge and of mind-body relations. It is described in the *Monadology* as 'the passing condition which involves and represents multiplicity in the unity [*unité*]'.[35] Monads are defined, then, by their capacity to perceive or express the multiplicity. Perceptions are not always conscious according to Leibniz, who calls conscious perceptions 'apperceptions'.[36]

Perception is recurrently described by Leibniz through two key visual metaphors: as a 'point of view' on the universe or its 'living mirror'. The 'point of view' of the monad or simple substance is what makes it different from all other monads, just as it is impossible to find in nature two leaves that are completely the same in terms of shape, colour, etc.[37] In other words, the singularity or individuality of the monad is constituted through its point of view. In Leibniz's metaphysics, the whole (universe) is defined by the part (monad) as much as the parts are determined by the whole. To say that every monad expresses the totality of monads is equivalent to saying that each monad contains the entire universe. This is possible if we understand the expression as a kind of representation. The entire universe is represented in every monad, just as a territory is represented by a map.

As mentioned, human minds are rational monads. The expression of the universe by rational monads is also described as 'perception'. Every monad expresses the world in its entirety, as mentioned, but they differ from one another in the degree of distinctness of their perceptions. This degree is also described as the 'point of view' of the monad.

The logic of expression guides Leibniz's theory of knowledge as well. Thus, he defines the 'idea' as that which expresses, within the human soul, a set of relations that correspond to the thing expressed.[38] In his short exposition 'What Is an Idea?' (1678), Leibniz describes the idea as 'something in me *which not merely leads me to a thing but also expresses it*'.[39] The set of relations in the idea which expresses a thing corresponds to that within the thing expressed, he writes. Ideas originate in God, who has impressed upon us a faculty enabling us to derive these sets of relations that correspond to those that are in things. Thus our idea of a circle expresses the circle itself by corresponding to it in a manner which enables us to derive from it truths about the circle.[40] In this way Leibniz creates a continuity and interdependence not only between all creatures and the universe but also between human perception and the external world. This inclusion or immanence of the world within human consciousness is part of what attracts Deleuze and Benjamin to the conception of expression.

Expression and continuity are linked in Leibniz's metaphysics through the active component of monads, which makes them 'living mirrors of the universe' – thus through the temporal aspect of expression.[41] Leibniz calls this self-active element in monads 'appetition'. The appetition is the 'living' element that drives the monad to actualize or realize itself; it can also be described as a drive or desire. Perception, on the other hand, is what constantly represents the universe.

The continuity between the expresser and the expressed may create, at its most extreme, a kind of 'mobius effect' in which it is no longer possible to distinguish interior from exterior.[42] This may be the case for Spinoza, but it is certainly not so for Leibniz, according to whom there is an isomorphic, hierarchical relation between the expresser and the expressed. Leibniz insists, notably, upon the monads' separateness from one another, their 'having no windows'.[43] The conception of expression enables, then, a fragile balance between the continuity of the monads with the world they express and their being closed off, or discontinuous from the world.

The balance is also retained through the introduction of 'pre-established harmony' [*une harmonie préétablie/prästabilierte Harmonie*]. According to this concept, one of the most important in Leibniz's system, upon creation, the development of each monad was pre-arranged to correspond to that of the changes in all other monads. Thus, the series of changes taking place in each and every monad harmonizes with those taking place in all other monads.[44] There is a sense, therefore, in which Leibniz's principle of continuity may be considered as historical, insofar as it concerns not only the temporal changes of every single monad but also how these affect the continuum of the totality of monads. It determines the way single elements of time correspond to time as a totality. Benjamin will refer to this as 'virtual history' in the 'Epistemo-Critical Foreword' to *Origin of the German Trauerspiel*.[45]

iv Perfection

While sparing them the impositions of Kant's transcendental idealism, Leibniz's structure of expression forced upon Benjamin and Deleuze a different problematic, which follows from Leibniz's principle of 'pre-established harmony', according to which all infinite series of monads are harmonized upon creation according to a divine plan or script. Leibniz illustrates this conception by describing several bands of musicians playing in separate rooms so that they do not see or hear one another, yet because each follows their notes, their music harmonizes perfectly.[46] The notes, determined by God, allow the musicians, or monads, to move in independent paths, not directly affecting one another yet remaining in tune. Elsewhere Leibniz describes harmony as created through the ideal influence one monad has on another.[47] As mentioned, monads are closed off from one another, yet they can affect one another on an ideal level.

What has been taken as Leibniz's 'optimism' is the idea that we live in the most perfect, most harmonious of possible worlds selected by God. Yet the concept of 'harmony' is in fact defined by Leibniz as 'unity within variety'. The main criteria for level of 'perfection' are a combination of simplicity and variety; as Leibniz describes in 'Principles of Nature and Grace':

> It follows from the supreme perfection of God that he chose the best possible plan in producing the universe, a plan in which there is the greatest variety together with the greatest order. The most carefully used plot of ground, place and time; the greatest effect produced by the simplest means.[48]

Leibniz describes the perfection of the world in economic terms, as the result of the greatest variety produced from the simplest or most orderly means. As presented in the *Theodicy*, for example, Leibniz's system is hierarchal and directed towards what he describes as the 'perfect' and the 'morally good'. All monads constantly strive to perfection, motivated by their 'appetition', which drives them from less perfect to more perfect perceptions. Leibniz's 'appetition', or appetite, resembles Spinoza's *conatus*, a directional force or drive that motivates the soul, whether consciously or unconsciously.

When describing the perfection of the soul, Leibniz writes that it 'has perfection to the extent that it has distinct perceptions'; that is, the soul becomes more perfect when its perceptions are less confused and closer to God's infinite knowledge. Though always strived for, perfection is unattainable by all beings other than God, which in a sense gainsays the optimistic reading of Leibniz's metaphysics, since humans are portrayed as constantly pursuing an unattainable goal.[49]

Moreover, despite the apparent unidirectionality of the monad's movement towards perfection, Leibniz's concept of pre-established harmony is at the same time inherently circular, since, as mentioned, the monad enfolds not only its future; rather Leibniz describes past, present and future as jointly embedded in every monad. This seems to suggest a non-linear understanding of time according to which the past is susceptible to changes as much as the present and future. In a similar vein, the notion of 'apokatastasis', or universal restitution, which Leibniz develops in several late texts, contradicts a conception supporting the unilateral directionality of history.[50] As opposed to a vision of history progressing continually towards perfection, these texts describe its eternal return. As we will see in Chapter 3, both Deleuze and Benjamin drew in different ways on these notions of circular, enfolded or nested temporalities as well as on Leibniz's understanding of eternal return.

Pre-established harmony is not a mere given, then, but a continual, repeated process of actualization, following God's selection of the best possible world. In this way, Leibniz also justifies his theory of 'natural retribution', according to which sins and virtues carry their own rewards or punishments.[51] The natural result of sin is punishment, just as nature is ordered to achieve the greatest possible order and balance, yet in both cases the order is not given or determined in advance but repeatedly selected by God.

The notion of harmony also guided Leibniz's understanding of space, time and motion.[52] In his later work, he developed conceptions of space and time as ideal, abstract orders that arrange the relations between phenomena. Space is viewed as the order of co-existence; time as the order of succession.[53] Time is therefore distinct from duration, as space is distinguished from co-location.[54]

Leibniz's denial that time and space are absolute, rather describing them as relational orders, has been the subject of much critical debate, since this kind of relational understanding of time and space would seem to pose challenges for its serving as a foundation for physics.[55]

The same logic that guides Leibniz's conception of expression determined his understanding of time. Thus, Leibniz defines the nature of being itself as temporal. Schepers argues that Leibniz's theory of time is modern insofar as, rather than substances merely being actors within space and time, they constitute the orders of space and time through their self-action. He notes the following example:

> [Leibniz] discovered the far-reaching consequences of the little word 'now' [*nunc*]. He explained that each time we say 'now', we actually refer to the whole world and the whole of time. Moreover, when we utter a factual sentence containing the little word 'is', we include the entire world (by saying 'It is cold', we mean 'it is cold here and now').[56]

In a similar way, Benjamin's concept of 'now time' [*Jetztzeit*] denotes the way in which past epochs are charged intensively with the forces of the present. 'Now time, as a model of messianic time, comprises the entire history of mankind in a tremendous abbreviation.'[57] The abbreviation enacted through 'now time' includes virtual history, both past and future. The ratio between 'now time' and the history of mankind, writes Benjamin, corresponds to that between the history of mankind and the history of the universe.[58]

3 A forbidden tradition

i Expression and immanence

It seems odd that Spinoza's philosophy, and not Leibniz's, is the focus of Deleuze's study on expression.[59] Expression as a problematic belongs more to Leibniz's philosophy, in its constant wrestling with the question of continuity, than it does to Spinoza's. Indeed, Spinoza seldom uses the term, and does not supply a definition for it, as Deleuze concedes.[60]

Deleuze writes that 'immanence is revealed as expressive, and expression as immanent'.[61] The original title of *Expressionism in Philosophy: Spinoza*, 'Spinoza et le problème de l'expression', clearly states that the problem of expression is its central one, while this gets lost in the English version of the title that emphasizes 'expressionism', a term with very different connotations, over 'expression'.

In fact, as becomes most clear in the final chapter of the book, in which Leibniz and Spinoza's systems are juxtaposed, Deleuze uses a problematic structure found in Leibniz in order to define Spinoza's philosophy as a philosophy of continuity and immanence, which lacks the hierarchical aspect of Leibniz's philosophy that arises out of his positioning of God outside his system. Through their conceptions of expression, Deleuze suggests, both Spinoza and Leibniz seek to inaugurate a continuity between the intelligible and the sensible, as well as a continuity between nature and history. These are shaped against Descartes' dualistic thought:

> Independently of one another the two philosophers seem to rely on the idea of expression in order to overcome difficulties in Cartesianism, to restore a philosophy of Nature, and even to incorporate Cartesian results in systems thoroughly hostile to Descartes's vision of the world. To the extent that we may speak of the Anticartesianism of Leibniz and Spinoza, such Anticartesianism is grounded in the idea of expression.[62]

The structure of expression is used by both philosophers in order to overcome Cartesian dualism and propose an immanent structure of thought. It is through developing this structure that they advance beyond Descartes' doctrine of 'clear and distinct' ideas, putting forward more complex epistemological accounts. As such, the structure of expression offers an alternative to the tradition of modern Western philosophy that evolved from Descartes.

Despite their shared critique of Descartes, Leibniz emerges as Spinoza's privileged opponent throughout *Expressionism in Philosophy*. Leibniz's philosophy is described by Deleuze as exemplary of an alternative version of 'the problem of expression' and its solutions. In contrast to Spinoza's designation of diverse attributes and modes as expressing a unity of being, in Leibniz, as aforementioned, each monad, as a unity, expresses the totality of monads. The problem is thus reversed, the main issue being the capacity of unity to express a multiplicity. As Joughin notes in his preface to the English edition, Spinoza and Leibniz offer

> two systems of folding: Spinoza's unfolded from the bare 'simplicity' of the infinity into which all things are ultimately folded up, as into a universal map that folds back into a single point; while Leibniz starts with the infinite points in that map, each of which enfolds within its infinitely 'complex' identity all its relations with all other such points.[63]

Both use the term 'expression' in order to define the interdependence between God, the universe, and the creatures and phenomena that inhabit it, as well as the interdependence between soul and body. The conception of expression described in *Expressionism in Philosophy* denotes a triadic relation between the expresser, the expressed and expression itself. The 'logic of expression', developed through a long tradition that runs from the Stoics to the Middle Ages, distinguishes between the sense of a proposition, which has no existence outside of itself, on the one hand, and what is designated by the proposition, the object or thing as such, on the other.[64] Here we encounter a significant characterization of expression: the existence of the expressed is contingent upon its expression by the expresser.

Deleuze describes the philosophical history of the conception of expression as 'rather hidden', and 'rather cursed [*maudite*]', since its proponents faced charges of pantheism.[65] Its immanent model poses a threat to emanative and creationist theological theories championing a transcendent God. In other words, since what is expressed has no existence outside its expression, the concept of expression is a 'specifically philosophical concept of immanence'.[66] The 'danger' is that what is expressed would be 'swallowed' by its expresser, or in its expression, and that nothing would remain outside of it.

As opposed to structures of transcendence, the structure of expression describes containment, a remaining within. This immanent structure implies that divine perfection is sought internally rather than exteriorized; therefore, expression encapsulates a 'conatus', or volition in a constant state of striving for perfection. These anti-transcendent tendencies of Baroque theology were similarly highlighted by Benjamin: 'There is no Baroque eschatology; and precisely for that reason there is a mechanism that multiplies and exalts everything earth-born before it is delivered over to its end.' Lack of hope for redemption in the next world drives finite beings to reach extreme modes of being before their end. The deprivation of an afterlife leads inevitably to a widening of the gap between the finitude of humans and the infinitude of God and the world famously epitomized by Pascal.[67]

The structure of expression entails ontological, epistemological and historical inclusion. According to the first, in fact an onto-theological immanence, the entire universe and God are included within every individual substance. Each monad contains the world so that the latter is immanent within the former. Universe and monad are interdependent; the one cannot exist without the other. The relation of expression exists not only between substance and universe; substances can express one another, and express God too (although in this case God is not reducible to the substances).

Historical immanence is described by Leibniz, for example, in his renowned formulation in the *Monadology*: 'every present state of a simple substance is naturally a consequence of the preceding state, in such a way that its present is pregnant with the future [*tellement que le présent y est gros de l'avenir*].'[68] Not only does the universe, made up of an infinity of monads, lie immanent within every individual monad but so too do its future, present and past, as Leibniz writes in 'First Truths'.[69] This type of inclusion is logical/ontological as well, since Leibniz defines every substance as including all its predicates. This last, ontological immanence was one of the main sources of interest in Leibniz for Deleuze, while historical immanence fascinated Benjamin.

ii Name, mirror and seed

The term 'expression' has two 'correlates', suggests Deleuze, namely *explicare* and *involvare*. Thus, the expresser both explicates and involves that which it expresses; the expressed is clarified and unfolded, yet at the same time remains folded, or implicated within the expresser: 'Expression in general involves and implicates what it expresses while explicating and evolving it.'[70] Within the philosophical-theological tradition of expression, in which both Leibniz and Spinoza partake, the 'one' or 'unity' expressed is that of God-nature. The notion of expression as manifestation of God's divine nature is also linked to the tradition of divine naming, to which the conceptions of expression of Leibniz, Spinoza and Benjamin refer. Benjamin writes, for example, that 'the theory of proper names is the theory of the frontier between finite and infinite language', suggesting that although the human name is expressed in finite language, it is also a unique expression of God's infinity.[71]

Expression indicates a relation of constant correspondence, either between two things or between two series of things, which is different from that of causality.[72] Deleuze names two metaphors that have been traditionally used to illustrate the conception of expression: that of the mirror and that of the seed [*le germe*]. In the first metaphor, the mirror stands for the expresser, while the expressed is 'mirrored', namely, that which is reflected. In the second, the relation between the seed and the tree is likened to the expressed (tree) and that within which it is expressed, in other words, the expression (seed). The first metaphor is spatial, and indicates a mimetic resemblance between the expresser and the expressed, a reproduction of the second in the first occurring in space. The second metaphor is temporal, implying a correspondence between the expressed and the expresser that Deleuze calls 'genetic'; the seed includes the genetic data of the tree, and indicates development or constitution over time.[73] We will see in Chapter 5 that the mirror and the seed reappear in Deleuze's concept of the 'crystal-image', designating two types of cinematic expression.

Within infinitesimal calculus, difference between quantities is understood through a common ground, or system, that views them as variables.[74] Leibniz's differential calculus formed the basis of his relational metaphysics, in which things are comprehended not by their stable essence but rather through their function in a system and their relations to other components. This relational approach and the differential relations that constitute it are key to Deleuze's reading of Leibniz both in *Difference and Repetition* and in *The Fold*, where, as we will see in Chapter 4, Deleuze describes this anti-essential, *eventual* metaphysics and ontology as 'Mannerist'.

iii Difference in itself

Leibniz's conception of expression plays an important, underestimated role in the project Deleuze sets out for himself in *Difference and Repetition*, namely, putting forth a philosophy of difference that would serve as an alternative to the dominant philosophy of representation.[75] 'Representation is the site of transcendental illusion', writes Deleuze.[76] He attempts to liberate 'thought from those images which imprison it', or to introduce a manner of thinking and philosophizing that is not curbed by the dogmatic assumptions that have been considered fundamental for philosophers to date. All previous attempts at a philosophy of difference, starting with that of Aristotle, subordinate difference to identity and representation, he writes.[77] While Leibniz is no exception in this regard, Deleuze finds in Leibniz's metaphysics a concept of 'difference in itself' that lies at the heart of the theory of sensibility he puts forward in *Difference and Repetition*.

In order to challenge the philosophy of representation, Deleuze must reach beyond the conception of language as a primary form of representation, and he does so by describing the idea as a mathematical relation. He compares the differential relation, symbolized as *dx*, with the relation of contradiction that dominates philosophy of representation – *non-A*. As he writes, '*dx* is the idea – the Platonic, Kantian or Leibnizian idea – the "problem" and its being'.[78] The differential relation is described as both the 'idea' and the logical or mathematical 'problem', two foundational concepts of Western philosophy which Deleuze defines as inseparable from one another, as we will discuss below, thus challenging the widespread definitions of each.

Differential calculus and its assimilation and reformulation within Leibniz's metaphysics are significant, then, for Deleuze's outline for a model of learning that challenges dominant philosophical models, specifically those of Kant, Hegel and Descartes. In Leibniz's infinitesimal system differences tend towards their disappearance or cancellation. Differential relations between terms constitute the terms themselves, as proved by Leibniz by showing that even when the terms 'vanish', the relation between them remains constant.[79]

Deleuze's appropriation of Leibniz's differential calculus is coupled with a strong critique of Leibniz's metaphysics that is softened in the later *The Fold*. In their converging towards God, Deleuze writes in *Difference and Repetition*, Leibniz's series tends towards a limit.[80] Deleuze criticizes Leibniz for the fact that, rather than diverging, his series always converge.[81] In other words, his critique is aimed at the hierarchal and continuous nature of Leibniz's system. The convergence of series in Leibniz's calculus as well as in his metaphysics means, for Deleuze, that Leibniz's differential is contingent upon identity. The only way difference could be freed from limitation would be if it affirms itself through repetition, thereby submitting to the law of eternal return.[82]

As Smith describes, infinitesimal calculus was derided, during the eighteenth and nineteenth centuries, for the 'dynamic notions' it supported, such as 'infinitesimals, fluxions and fluents, thresholds, passages to the limit, continuous variations – all of which presumed a geometrical conception of the continuum'.[83] These dynamic notions are precisely what attracted Deleuze, and perhaps Benjamin as well, to Leibniz's differential calculus. Deleuze was well aware of developments in the history of mathematics that contradict the geometrical principle, central to Leibniz's calculus, of 'approaching a limit'. This principle was replaced in the late nineteenth century, by the efforts of mathematicians such as Karl Weierstrass and Georg Cantor, with an arithmetic understanding of continuity.[84] Despite these developments, differential calculus remained crucial for Deleuze's philosophy of difference.[85]

4 Idea as monad

i Philosophical presentation

In the Foreword to *Origin of the German Trauerspiel*, Benjamin critiques subject-based conceptions of knowledge [*Erkenntnis*], as well as conceptions of history and historiography that are founded upon these. He seeks to inaugurate a new 'philosophical history' [*philosophische Geschichte*] rooted in what he calls 'natural history' [*natürliche Geschichte*]. This philosophical history, as knowledge and practice, would concern the presentation of truth, rather than serving as a simple guide to knowledge acquisition. Benjamin distinguishes between the domains of knowledge and truth, the latter alone being the concern of philosophical history.

According to Beatrice Hanssen, the term '*erkenntniskritisch*' in the title of the 'Epistemo-Critical' Foreword should be understood literally, as a 'radical repudiation of epistemology and transcendental philosophy' rather than a critique of the condition of possibility for cognition.[86] Benjamin's Foreword consists of a critique of the philosophical tradition of *Erkenntnistheorie*, or theory of knowledge, most prominently represented by Kant, and the reasonable [*vernünftig*] subject it presupposes. He continues therein a path he inaugurated in several fragments written between 1916 and 1919, culminating in the essay 'On the Program of the Coming Philosophy' (1918), in which a scathing critique is directed towards Kant and neo-Kantian philosophers.[87]

The conception of expression is present already in the epigraph to Benjamin's Foreword, drawn from Goethe's Materials to *History of the Theory of Colors*:

> Since no whole can be brought together in either knowledge or reflection, seeing that the former lacks internality and the latter externality, we must necessarily think of science as art if we are to expect from it any sort of wholeness. And it is not in the general, in the boundless, that we should look for this, but, just as art is always wholly present [*darstellt*] in each individual artwork, so should science always be wholly manifest in each particular matter treated [*in jedem einzelnen Behandelten*].[88]

Famously a disciple of Spinoza, Goethe describes the process of reflection and the gaining of scientific knowledge through a relation of expression between the detail and the whole. The above corresponds with either Spinoza or Leibniz's conceptions of expression, in which a totality is condensed into singular modes or monads. It is telling that Benjamin chooses this citation, which highlights the significance of the particular over the general, as an epigraph to the Foreword. The Foreword traces the philosophical history of the German mourning play, or *Trauerspiel*. Yet before doing so, it calls into question the basic assumptions of these domains by critically discussing the production and acquisition of knowledge. Goethe's citation suggests that science [the German *Wissenschaft*, literally created-knowledge] should be thought of as art – just as art as a whole is present in every single artwork, so scientific knowledge as a whole is represented in every tiny detail of nature. Benjamin's Foreword deals primarily with the presentation [*Darstellung*] of knowledge and history. These, he argues, are manifest as totalities in ideas of art. From the epigraph and on, Benjamin designates their presentation in language as expressive in nature. Years later, Benjamin returned to Goethe in the *Arcades Project*, referring to his notion of the *Urphänomen*, which denotes the appearance of ideal essences within sensual forms.[89]

The detail or particular, in which science reveals itself as complete, has a privileged place in the epigraph. The Foreword unfolds a philosophic-historical theory of interpretation of art, that, on the one hand, considers specific artworks, in this case plays, in relation to the *Trauerspiel*, which Benjamin defines as an 'idea'. On the other hand, the plays will be considered as wholes in relation to the internal details of their composition. One of Benjamin's overall arguments in the book relates to the originality of these German mourning plays, which lies in fine details rather than in the plays viewed as complete artworks.[90]

Benjamin uses both *Darstellung* (presentation) and *Ausdruck* (expression) in the Foreword; the two terms, however, are not interchangeable. Rather, *Darstellung* denotes, for Benjamin, specifically presentation in language, and consists of a type of expression. A third term Benjamin uses in texts written during the same years in which he worked on the *Trauerspiel* book, which may be understood through Leibniz's conception of expression, is *Verwandtschaft*, which is used to describe a non-casual, non-visual correspondence or affinity.[91] Benjamin develops a theory of such correspondences in the texts 'Doctrine of the Similar' and 'On the Mimetic Faculty', written in 1933, in which he discusses the human capacity to perceive similarities and 'natural correspondences' that serve to illuminate occult knowledge.[92]

As opposed to mathematical formulations, language discloses philosophical truths by encompassing the tension between expression and the expressionless. 'Philosophical teaching' [*Lehre*] is 'based on historical codification'. Benjamin writes in the Foreword, it 'cannot therefore be evoked *more geometrico*'.[93] Spinoza's *Ethics*, the archetypal philosophical system presented in geometrical-like deduction, is implied by the term *more geometrico*. Philosophical teaching must not be abstracted into mathematical propositions as attempted by Spinoza, Benjamin suggests; rather, it must be formulated in language that is embedded in history.

For Benjamin, the question of philosophical presentation [*Darstellung*] concerns the relationship between the singular and the general, as he illustrates by comparing philosophical presentation to piecing together a mosaic:

> Just as the majesty of mosaics remains intact when they are disassembled into capricious bits, so philosophical observation fears no dissipation of momentum.

> Both come together out of the singular and disparate; nothing could attest more powerfully to the impact of what is transcendental – be it a saint's image or the truth. The value of thought-fragments [*Denkbruchstücken*] is all the more decisive the less they are immediately capable of measuring themselves by an underlying conception, and the brilliance of the presentation depends on this value to the same extent that the brilliance of the mosaic depends on the quality of the poured glass.[94]

The completeness of philosophical presentation, he contends, depends on fragments of thought and momentary observations, like a mosaic that comes together from disparate pieces that seem perhaps worthless on their own but together create a stunning image. The figure of the mosaic anticipates his later concept of montage, and together with the epigraph to the Foreword, attests to the expressive origins of montage as a concept that combines expression and construction.

Benjamin had studied, As Ng notes, Bernard Bolzano's *Paradoxien de Unendlischen* (1851), which provided a basis for Cantor's definition of the infinite set and contemplated the conception of the infinite himself, as we will see in the next chapter.[95] Yet, as opposed to Deleuze's recourse to differential calculus in an effort to counteract representation, Benjamin appropriates the mathematical infinite into a theory of presentation in language that prioritizes the singular, micrological and infinitesimal over the systematic and the general.

ii The idea is a monad

In the Foreword, Benjamin describes philosophical teaching as an 'order of ideas', informed by Plato's doctrine of ideas and Leibniz's metaphysics. He thereby extends the *erkenntniskritisch* project he had commenced in the earlier texts discussed, yet with some differences. Truth is contrasted with knowledge and is accessed by means of philosophical history rather than religion. Philosophical history, rather than religion, is also the place where ideas are formed.

> The idea is something linguistic and, indeed, in the essence of the word it is in each case that moment in which the word is symbol. In empirical hearing, in which words have been disfigured, there attaches to them, along with their more or less hidden symbolic side, an overt profane meaning. It is the concern of the philosopher, through presentation, to reestablish in its primacy the symbolic character of the word, wherein the idea comes to a self-understanding that is the opposite of all outwardly directed communication.[96]

In 'On the Program of the Coming Philosophy', Benjamin had argued that philosophical knowledge must be expressed in language rather than mathematical formulas. Here he similarly designates the idea as that through which, by its representation in language, true knowledge may be restored, as opposed to the kind of knowledge sought by epistemological theories that view knowledge as something that one can possess and learning as a process of acquisition.

It is philosophy, rather than the hitherto practised literary-historical analysis, that is necessary for the insight into the 'origin' of German *Trauerspiel*. In Leibniz's theory of expression, Benjamin identifies a possibility of designating ideas as a unity holding a multiplicity. Benjamin criticizes the proponents of literary-historical analysis for attempting to apply scientific procedures to art by using inductive or deductive methods. These methods, he writes, attempt to deduce the essence of artworks through external methods. Benjamin describes the *Trauerspiel* itself as an idea, rather than as a 'concept' or 'genre', and this enables him to view every single play as expressing a totality. He targets Benedetto Croce for favouring systems of classification that prioritize dividing artworks into genres and kinds according to external characteristics, over historical analysis that would seek their singular essence and origin.[97]

'The idea is a monad: the representation of phenomena rests preestablished in it, as in their objective interpretation [*objective Interpretation*]', writes Benjamin in the section of the Foreword titled '*Monadology*'.[98] As an expressive centre, the idea mirrors the pre-established relations between the phenomena of the world, and this mirroring, constantly in flux to reflect the changing world, consists of their 'objective interpretation'. This capability of expressing the totality of the relations between phenomena that make up the world affords the idea of access to 'true scientific content'.[99] The task Benjamin sets for himself in *Origin of the German Trauerspiel* is to offer such an 'objective interpretation' of Baroque plays. To this end, he follows Leibniz, whom he describes as a philosopher who penetrated so deeply into the real as to reveal an 'objective interpretation' of the world.[100]

The notion of 'objective interpretation' is not synonymous with the conception of 'immanent critique' Benjamin appropriates elsewhere from the early Romantics. Rather, here Benjamin strives to go beyond the Romantics' understanding of critique.[101] To be sure, 'objective interpretation' is immanent in the sense that, unlike the deductive or inductive methods of interpretation, it is internal to the artwork, dependent on neither recipient nor critic. Yet more important for the 'objectivity' in question is the 'idea' as an intermediary category between the singular artwork and art as a whole, and its totalizing effect. The interpretation of each artwork is immanent to the idea because the idea expresses the world as a totality. Through its totalization the idea is 'redeemed'.[102]

As expressive centres, ideas cannot be visually apprehended. This is because ideas, like monads, consist of a structure of relations, in which the whole is mirrored in the part from a certain point of view. Benjamin describes the idea as a linguistic component: every word has a profane meaning, on the one hand, and a non-communicable 'symbolic essence', the idea, on the other. This is the 'historical codification' to which Benjamin referred earlier, described in this case as the act of naming. Benjamin defines the idea through its temporality; the idea becomes such in the moment in which the word becomes symbolic.[103]

Through representation in language, philosophical contemplation can restore these symbolic essences or ideas, by bringing ideas to 'self-understanding' [*Selbstverständigung*], an independent understanding of their internal nexus of relations, that is the opposite of mediation/communication [*Mitteilung*]. The harmonious relation between ideas, each an isolated essence, constitutes truth.

Designating the idea as expressive centre/symbolic essence affords Benjamin the following. First, it allows an internal, objective, interpretation of ideas, through 'self-understanding'. Objectivity may be reached by deep immersion in reality, as Benjamin learns from the Baroque, which he views as an age of 'monstrous depth'.[104] Second, it enables Benjamin to precisely define the connection between language and history that lies at the basis of his philosophical history. From Benjamin's critique of the Romantics' 'speculations' in which 'truth assumed the character of a reflective consciousness', we gather the importance for him of the 'linguistic character' of truth, its inclusion of an image of total history and its consisting of a structure of relations between independent essences, which Benjamin likens to 'suns' or planets, each revolving in its own orbit.[105] What is symbolized or expressed by the idea, then, is the world including its history, not God or the absolute. Perhaps for this reason Benjamin models his theory of ideas on Leibniz's expression rather than Spinoza's.

As discussed, Deleuze highlights the fact that in Leibniz's thought, metaphysics and ontology are grounded in mathematics.[106] Deleuze targets Plato's conception of 'idea' and Kant's 'transcendental idea', criticizing the conception of the 'innate idea', and innateness more generally, as an *a priori* regulatory idea, and challenging the conception of knowledge as essential. Deleuze's own conception of the 'problematic idea', according to which the idea is redefined as problem, is indebted to Leibniz's understanding of knowledge and the manner in which it was informed by infinitesimal calculus.

Redefining the idea as 'problem', Deleuze evokes the distinction, developed within the history of mathematics, between problems on the one hand and axioms or propositions on the other. While 'axioms' define objects through their stable essence and derived properties, 'problems' designate objects by their capacity to be affected.[107] Criticizing Aristotle and Descartes, Deleuze argues that these thinkers reduced 'problems' into 'axioms' judging a problem to be 'true' according to the possibility of its solution. Leibniz, by contrast, was aware of the profound gap between propositions and problems or questions, writes Deleuze, especially those questions that investigate the circumstances of things.[108]

Deleuze refers to Leibniz's *New Essays on Human Understanding*, Leibniz's book-length rebuttal of John Locke's *An Essay Concerning Human Understanding*, in which Leibniz also unpacks his own theory of knowledge. Therein, Leibniz defines questions that 'ask how and ask for details' as a midway category between propositions and ideas.[109] In addition, Deleuze's theory of problematic ideas is indebted to Leibniz's theory of knowledge in its understanding of the process of learning as an actualization. Deleuze describes problem-solving as the engendering of 'real discontinuity' from 'ideal continuity', through a process of actualization. Problems/ideas harbour virtual relations that are incarnated in the actual relations defined by the 'field of solutions'.[110] 'Problematic ideas are not simple essences, but complexes, multiplicities of relations and corresponding singularities', writes Deleuze.[111] His description of the idea as a complex of relations is close to Leibniz's definition of the idea as that which expresses, within the human soul, a set of relations that correspond to the thing expressed.[112]

iii Method is indirection

Deleuze and Benjamin both criticize Kant's theory of experience and knowledge, by recourse to Plato's theory of ideas and Leibniz's monad and differential calculus. Yet as opposed to Deleuze's turn to differential calculus to counteract representation, Benjamin appropriates the mathematical infinite into a theory of presentation in language.

Benjamin's detailed engagement with Leibniz's conception of 'expression' commences in *Origin of the German Trauerspiel*, yet we have seen that some of his earlier written texts and fragments, in which he argues for a continuity between experience and knowledge in opposition to Kant's theory of experience, also make clear the stakes of Benjamin's early reading of Leibniz. These concerns underlie what we understand as a theory of learning, and of learning to philosophize, offered in the Foreword. Therein Benjamin makes use of the post-Kantian conception of 'infinite task', a term rooted in Leibniz's 'infinite analysis'.

This conceptual network, in my view, lies in the background of Benjamin and Deleuze's critical epistemologies. At the heart of their theories of learning is an abolition of the subject-object divide, as well as a challenge to disciplinary boundaries, specifically those between philosophy, art and science.[113] Both thinkers, moreover, draw a direct link between philosophizing and learning more generally. For Deleuze, philosophy consists of a 'pedagogy of the concept',[114] while Benjamin describes philosophical history as an apprenticeship of the mind.[115]

At the same time, both critique Leibniz's notion of the teleological system and propose. Conversely, philosophy involves digression or deviation from a linear path. Unlike Leibniz's image of the labyrinth, which implies an idea of philosophy that has the single aim of solving a puzzle or exiting the labyrinth, according to Deleuze, 'it would not be false to consider that philosophy is in a perpetual state of digression or digressiveness'.[116] In a comparable manner, Benjamin writes:

> Method is indirection … In its persevering, thinking constantly begins anew; with its sense of the circumstantial, it goes back to the thing itself. This continual breathing in and out is the form of existence most proper to contemplation.[117]

Philosophizing and thinking are roundabout, indirect processes. Often interrupted, they are continually forced to begin anew. These disruptions are part of the organic nature of thinking. What's more, as we will see in detail in Chapter 4, all three thinkers emphasize learning and philosophizing as embodied practices. While Leibniz compares reasoning to walking through a labyrinth, Benjamin uses an even more deeply embodied metaphor of breathing to describe contemplation.

2

Infinite tasks of learning: The Baroque-inspired epistemologies of Benjamin and Deleuze

Introduction: Open education

In the previous chapter we encountered Benjamin's redefinition of the idea as an expressive monad, and his understanding of philosophizing as an embodied process that necessitates roundabout, indirect paths. This chapter will focus on Benjamin and Deleuze's concepts of learning, which both relate, in different ways, to Leibniz's infinitesimal calculus and to the neo-Kantian notion of an 'infinite task'. By highlighting this conceptual trajectory, the chapter explores how Leibniz's calculus informed his own and later philosophies of knowledge, and more generally the relations between mathematics and philosophy, and how it informed Benjamin and Deleuze. What we term Benjamin and Deleuze's 'critical epistemologies' (in the case of Deleuze these terms must be used cautiously) serve as a foundation for their philosophies of history, politics and media.

Themes of education and learning run through both authors' works. Deleuze also taught philosophy, and reflected upon teaching, throughout his career. In both cases, these discussions are grounded in a reconsideration of the nature of knowledge itself. Their respective redefinitions of processes of learning as independent of the subject were informed by Leibniz, as we will see. Deleuze and Benjamin's reconfigurations of the relations between knowledge and experience serve as grounds for their alternatives to key modern epistemological theories, especially Kant's.

The chapter presupposes that Deleuze and Benjamin's 'returns' to Leibniz replicate, to some extent, those of Kant critics Cohen and Maimon, and we will therefore briefly examine these thinkers' turn to Leibniz in order to redefine Kant's conception of experience. Benjamin and Deleuze were informed by these interpretations and yet diverged from them too. Following Maimon, Deleuze finds in Leibniz a possibility of conceptualizing cognition as based on eternal genesis and not external conditioning, as in Kant. This emphasis on the internal generation of knowledge is shared in part by Cohen. Benjamin is critical of these ideas, which, he believes, highlights 'logical experience' over transient experience, that is, experience that is embedded in history.

Challenging Kant and Descartes' epistemologies, Benjamin and Deleuze offer theories of learning, viewed as a dynamic and infinite process, rather than as the acquisition or possession of knowledge, informed by Leibniz's conception of 'infinite analysis', and the post-Kantian conception of 'infinite task'. They follow attempts made by Cohen, Ernst Cassirer and Maimon, who return to Leibniz in order to rectify Kant's transcendental system. Yet while Benjamin's concept of expression is based on an interpretation of Leibniz's monad as a limitation of the infinite, Deleuze calls for an unlimited expression.

While Benjamin's reception of Leibniz is mostly implicit, Deleuze overtly utilizes Leibniz's conceptions of expression and differential calculus, harnessing their subversive nature, within his theory of learning, which is constructed by way of a critique of Kant, Descartes and Plato's theories of ideas and knowledge. As we will see, Deleuze highlights Leibniz's 'minute perceptions' as a theory that concerns the body, developing from Leibniz's notion of the 'complete individual' an embodied theory of learning as an event. Moreover, there remains a wide gap between the function of these common themes within the larger context of Benjamin and Deleuze's works. For Benjamin, Leibniz's 'expression' serves to designate continuity between knowledge and experience, history and nature (Benjamin is renowned for his concepts of discontinuity but as we will see, this discontinuity is preconditioned on continuity). One cannot speak of continuity between experience and knowledge in Deleuze, since, in an attempt to outline an affective concept of learning, he dispenses with both terms.

1 Infinite analysis and infinite task

i 'Problematic ideas'

> *Saying that learning is, after all, an infinite task would be in vain [On a beau dire qu'apprendre, après tout, est un tâche infinie]: it is none the less cast with the circumstances and the acquisition of knowledge, outside the supposedly simple essence of knowledge in the form of an innate or a priori element, or even a regulative Idea.*[1]

There's no point in saying that learning is an infinite task, since it is often defined, in any case, by the external conditions of knowledge-acquisition rather than the internal essence of knowledge. Many mistake the empirical conditions of knowledge-production as learning, Deleuze writes. Yet treating learning as an infinite task does not go far enough, because although it implies open-ended, continuous learning, it does not define learning as a process of encountering what is completely new, different and unrecognizable.

Deleuze's redefinition of learning stems from his critique of what he calls, in the third chapter of *Difference and Repetition*, 'the image of thought'. This 'image' is the manifestation of dogmas that stand as obstacles to free, independent thinking. These obstacles take the form of common sense and universal assumptions, things we consider obvious, those things that 'everybody knows, no one can deny' – that produce dogmas.[2] It can also be described as a kind of freezing, or fixation of dynamic thought processes. While Deleuze describes this image as 'pre-philosophical', he also considers the entire Western philosophical tradition as based on this restrictive mode of thinking.

In this mode of thought, thinking is equated with the recognition of the identical, that which is known. The epistemologies of Descartes and Kant, for example, use this model of recognition, in which the thinking I remains identical to itself through the thought process. For Deleuze, the thinking process does not consist of recognizing what the thinker already knows; it is a completely dynamic process in which the thinker and that which is thought are transformed. The model of recognition constitutes, according to Deleuze, a 'hindrance to philosophy'.[3] In order to philosophize, another model is necessary. 'For the new – in other words difference – calls forth forces in thought which are not forces of recognition, today or tomorrow, but are the powers of a completely other model, from an unrecognised and unrecognisable *terra incognita*.'[4] The model of recognition, in other words, cannot be used in order to be able to learn new things that are different from what we already know.

Deleuze is implicitly referring here to what is often described as 'Meno's paradox', from Plato's dialogue *Meno*. The paradox addresses the possibility of inquiry, and is formulated by Socrates, in reply to Meno, as the following: inquiry is impossible because if one inquires after what one already knows, the inquiry is unnecessary because one already knows, nor can one inquire into what is unknown, because one would not know what to inquire into. Deleuze describes the model of recognition as describing inquiry into what is already known, or identified, which for Deleuze, cannot be considered as learning.

Deleuze embraces Leibniz's differential as part of his redefinition of learning that is independent of recognition, while rejecting its Kantian and neo-Kantian interpretation in the concept of the infinite task. Problems, he writes, are the 'differential elements in thought'. The differential, as defined by Leibniz, determines the infinitesimal differences between successive values of a variable.[5] Problems, in other words, are what generate internal differences in thought.

As part of this theory of learning Deleuze redefines not only the problem but also the 'idea', synthesizing the two epistemological categories in the concept of the 'Problematic Idea', which, Deleuze writes, was inspired by the Kantian method of Critique, that framed the Idea as problematic.[6] However, as Kant's Critique remained dominated by the dogmas, Deleuze sought to 'correct' it through the introduction of Leibniz's differential. Deleuze therefore grounds his theory of learning in the mathematical concept of the differential.[7] At the same time, he critiques the widespread alinement, in the Western Philosophical tradition, between problems on the one hand and axioms or propositions on the other, which ignores the distinction between them. As Deleuze highlights, while 'axioms' define objects through their stable essence and derived properties, 'problems' designate objects by their capacity to be affected.[8]

Criticizing Aristotle and Descartes, Deleuze argues that these thinkers reduced 'problems' into 'axioms', judging a problem to be 'true' according to the possibility of its solution. Leibniz, by contrast, was aware of the profound gap between 'propositions' and 'problems', writes Deleuze. While the first are defined by external events, by 'the how and the circumstances', the second are defined internally.[9] Deleuze is referring to Leibniz's *New Essays on Human Understanding*, in which Leibniz designates 'questions' that 'ask how and ask for details' as a midway category between propositions and ideas.[10] Leibniz suggests therein that knowledge should not be reduced to distinguishing between true and false, nor is it solely based on empirical experience. Ideas can be explained in a way that elucidates their sense without referring to their truth status, Leibniz says, in the same way in which commentaries explain the Holy Scripture.[11]

'Problematic Ideas are not simple essences, but complexes, multiplicities of relations and corresponding singularities', writes Deleuze.[12] As opposed to conceptions that view the idea as fixed, such as the Platonic idea, Deleuze's 'Problematic Ideas' are fluctuating and relational. His description of the idea as a complex of relations is close to Leibniz's definition of the idea as that which expresses, within the human soul, a set of relations that correspond to the thing expressed.[13]

Deleuze describes problem-solving, which forms part of learning, as the engendering of 'real discontinuity' from 'ideal continuity'.[14] As such, learning bears the structure of the virtual, which Deleuze also defines through the differential: 'the virtual is opposed not to the real but to the actual. *The virtual is fully real in so far as it is virtual* ... the reality of the virtual consists of the differential elements and relations along with the singular points which correspond to them.'[15] Rather than a realization of a possibility or a potentiality, learning something new and different can be described, for Deleuze, through the structure of the virtual.[16] In his concept of 'virtual', Deleuze draws not only on Leibniz's calculus but also on his concept of virtual, which will be discussed in the next chapter.[17]

Deleuze further elaborates his theory of learning by distinguishing between knowledge [*le savoir*], attained by apprenticeship [*l'apprentissage*], on the one hand, and learning [*l'apprendre*], on the other: 'learning is the appropriate name for the subjective acts carried out when one is confronted with the objectivity of a problem (Idea), whereas knowledge designates only the generality of concepts or the calm possession of enabling solutions.'[18] Unlike the 'apprenticeship' of the spirit in Hegel's *The Phenomenology of the Spirit*, in which learning is subordinated to absolute knowledge, Deleuze writes, in his own model there is no separation or hierarchical order between the process of learning and that which is learnt.

ii Philosophical lineage of the infinite task

The neo-Kantian concept of an 'infinite task', developed by philosophers such as Cohen and Cassirer, originates in Kant's discussions of the infinite, as well as Leibniz's conception of 'infinite analysis'.[19] Deleuze suggested, in one of his lectures given in Vincennes, that Leibniz's conception of 'infinite analysis' was a novel one in the history of philosophy.[20] Leibniz himself viewed this conception as one of his most significant contributions, since he believed it had solved the problem of enabling contingency in a world in which 'every predication has a foundation in the nature of things'.[21]

Leibniz's 'infinite analysis' describes the idea that there are true propositions in which the connection between subject and predicate may not be demonstrable in a finite number of steps. In other words, a proposition can be true yet not demonstrable.[22] The conception of infinite analysis applies the mathematical notion of asymptotes to conceptual analysis:

> Just as the smaller number is contained in the larger in every proportion … so in every truth the predicate is contained in the subject … But in proportions the analysis may sometimes be completed, so that we arrive at a common measure which is contained in both terms of the proportion an integral number of times, while sometimes the analysis can be continued to infinity, as when comparing a square with the diagonal. Just so, truths are sometimes demonstrable or necessary, and sometimes free and contingent, so that they cannot be reduced to identities as if to a common measure by analysis. This is the essential distinction between truths as well as proportions.[23]

From the conception of infinite analysis follows the conclusion that there are two kinds of truths: contingent truths, which are not demonstrable, and necessary truths, which one can demonstrate in a finite number of procedures. Only God has access to contingent truths, since he sees the series in its entirety, while human knowledge is always partial.

Leibniz's application of the mathematical infinite to conceptual analysis was significant to Hermann Cohen's development of his own conception of 'infinite task', in turn important for both Benjamin and Deleuze. In his reading of Leibniz, Cohen highlights Leibniz's designation of thought as continuous, acknowledging the manner in which Leibniz applied his principle of continuity and reflections on infinity more generally to domains other than mathematics. Cohen wrote:

> *Continuity became the higher concept, from which infinity derives* … For the *being* of infinity is now grounded in the *thought* of continuity. Continuity, in as much as it is *idea*, and in as much as it is *law*, is now a principle both of *reason* and of *nature*.[24]

To Cohen's mind, the significance of Leibniz's principle of continuity lies in its framing of relations between elements in nature and in thought through the infinitesimal method.

Cohen uses the conception of 'infinite task' primarily in ethical and theological contexts, where he defines the possibility of participating in God's holiness as an 'infinite task'.[25] In Cohen's system, this participation, or the 'correlation' between man and God, as he defines it, is executed through ethical action and is also the object of philosophical reason.[26] As in the case of Leibniz's infinite analysis, Cohen's infinite task relates to free will, yet in the case of Cohen the infinite task consists of an ethical-religious choice made by the individual, constituting the never-ending movement of the individual towards God.[27]

Deleuze compares Kant's synthesis of finitude with Leibniz's infinite analysis in a lecture from the Vincennes series.[28] While in Leibniz every analysis is conceptual, and therefore quantitative differences can be reduced to qualitative ones, writes Deleuze, for Kant knowing consists precisely in going beyond the concept. Kant's revolution in understanding space and time involved their irreducibility to the order of concepts. By returning to Leibniz, then, Deleuze reinstates conceptual difference, significant for his understanding of philosophy as the creation of concepts.

As he writes, Kant and the tradition that followed him discarded the infinite by subordinating it to the finite. The goal of reintroducing infinite analysis, for neo-Kantians like Cohen, is that of 'substituting the point of view of genesis for that of conditioning', in order to show that the conditions of appearance form the genetic elements of that which appears.[29] However, unlike Leibniz and Cohen's conceptions of the infinite, Deleuze's own concept of infinity does not subordinate itself to finitude. Infinity is rather constituted by exceeding finitude, engendering appearances through this act of surpassing the finite.[30]

iii Continuum of experience

We have seen in the previous chapter that Benjamin held that philosophical style, or the way philosophy is expressed in language, is of utmost importance. In 'On the Program of the Coming Philosophy', he writes:

> For Kant, the consciousness that philosophical knowledge [*philosophische Erkenntnis*] was absolutely certain and a-priori, the consciousness of that aspect of philosophy in which it is absolutely the peer of mathematics, ensured that he devoted almost no attention to the fact that all philosophical knowledge has its unique expression [*Ausdruck*] in language and not in formulas or numbers … [I]t is ultimately because of this fact that the systematic superiority of philosophy over all sciences as well as mathematics will be asserted.[31]

Benjamin charges Kant with viewing philosophical knowledge as entirely *a priori* and certain, and therefore disregarding its essentially linguistic nature. The fact that philosophy may only be expressed in language is precisely what differentiates it from mathematics, Benjamin argues, and furthermore, the reason for its superiority over all other sciences.

Written in 1917, and completed in 1918 with an addendum, 'On the Program of the Coming Philosophy' defines the 'central task of the coming philosophy' in relation to Kant's 'triadic system' or three Critiques: the *Critique of Pure Reason* (1781, 1787), the *Critique of Practical Reason* (1788) and the *Critique of the Power of Judgment* (1790). The 'coming philosophy', that is, the philosophy of the future, would necessitate a 'revision' of Kant's philosophy, Benjamin writes, in which some elements would be retained, and others rejected.

During the years preceding the writing of this text, as a student in Freiburg, and later in Berlin and Munich, Benjamin read works by Kant and neo-Kantian thinkers, and had studied with several Marburg School professors.[32] He also attended lectures by Cohen, as well as studying Cohen's *Kant's Theory of Experience*, as testified by Scholem.[33]

These were years during which Benjamin and Scholem engaged in a continuous intellectual dialogue, which informed Benjamin's thought in two main directions: first, Scholem introduced Benjamin to Judaic philosophical and Kabbalistic texts, and second, as has been the focus of research in recent years, Scholem had sparked Benjamin's interest in logic and mathematics.[34] Traces of these themes may be found in 'On the Program of the Coming Philosophy', which launches a double critique against Kant and neo-Kantianism.

Philosophy, as perceived in this text, is the philosophy of knowledge, *Erkenntnistheorie*.[35] The task of a future philosophy is therefore an epistemological one. Benjamin defines the double problem of *Erkenntnistheorie* as follows: 'First of all, there was the question of the certainty of knowledge that is lasting, and, second, there was the question of the integrity [*Dignität*] of experience [*Erfahrung*] that is ephemeral.'[36] Knowledge is not composed uniquely of eternal truths, according to Benjamin; it also comprises the temporal, fleeting experiences that serve as its immediate objects.[37]

Benjamin criticizes Kant for a conception of knowledge that prioritizes the certainty of *a priori*, timeless knowledge, reducing the concept of experience to that of 'primitive, self-evident experience' and ignoring experiences that occur in time and are therefore passing. He argues that in the *Prolegomena*, Kant sought the principles of experience in sciences, and that the post-Kantians pushed this tendency even further, completely identifying experience with the 'object-world' [*Gegenstandwelt*] of science.[38] According to Benjamin, this scientifically grounded conception of experience is reductive, and hence the philosophy of the future must adopt a 'higher' conception of experience that would include realms other than science, most importantly, that of religion.

Kant's narrow conception of experience, according to Benjamin, results from the binaries that dominate the internal structure of his system. First, Kant conceives of knowledge as a subject-object relation. Second, Benjamin regards the distinction between appearance and thing-in-itself, *noumenon* and *phenomenon* – and more generally the view according to which individuals receive sensations from external objects and form ideas based upon these – as an empirical residue within Kant's epistemology. And finally, there is the binary delineated by Kant between the intelligible and the sensible, which the neo-Kantians sought, more successfully, to transcend.

Although Kant did not mean to shatter experience into its singular domains, these binary structures ruptured the continuity of the concept of experience across various fields, resulting in its fragmentation. Benjamin proposes the formation of a 'pure and systematic continuum of experience' that would be 'multiply gradated' stretching across domains such as art, law, and history as well as the natural sciences.[39] That is, although forms of experience belonging to different domains, such as historical and artistic experience, are distinct from one another, it is a difference of degree, while the essential nature of experience remains the same. This unified nature, Benjamin writes, would manifest itself in ideas that converge towards a higher knowledge.[40]

The conception of experience as gradated, and as manifest in ideas that express a totality of knowledge, is close to the theologically oriented system of Leibniz, in which the imperfect ideas in human consciousness express God's perfect knowledge. Yet, Benjamin denies that any pre-Kantian philosopher had conceived of this epistemological task [*erkenntnistheoretisch Aufgabe*]. Kant's contribution to redefining experience was unprecedented, and hence the problem may only be defined as such from a post-Kantian perspective.[41] Yet at least one aspect of the problem, the question of how human consciousness is related to pure knowledge, may be outlined by a return to pre-Kantianism.[42]

The alternative structure of relations between experience and knowledge that Benjamin proposes is one in which experience and knowledge are interdependent. Not just experience but knowledge too must be transformed. The new conception of knowledge Benjamin proposes would be continuous with experience, serving as its teachings [*Lehre*].[43] In other words, knowledge as teaching, presented systematically in language, is described as the actualization of experience, its continuation and fulfilment. 'Experience is the uniform and continuous multiplicity of knowledge', Benjamin surmises.[44]

Benjamin reverses the traditionally empirical conception of the relations between experience and knowledge, according to which every experience is singular and therefore experience as a whole consists of a multiplicity, while knowledge is the unified generalization of experience. He argues, on the contrary, that experience is essentially unified, and that this unity reflects the multiplicity of knowledge.

iv The task that cannot be given

Although Benjamin does not explicitly define, in 'On the Program of the Coming Philosophy', philosophy's task as an 'infinite' one, the conception of 'infinite task' is present therein. During the period when this essay was written, Benjamin had decided to write his dissertation on the 'concept of the infinite task in Kant', which, had he written it, may have consisted of a reading of Kant in a neo-Kantian vein, informed by Cohen's *Logic of Pure Knowledge* [*Logic der Reinen Erkenntnis*].[45] Despite Benjamin's criticism of neo-Kantianism in 'On the Program', Cohen's understanding of knowledge as a process, in a state of constant becoming, manifestly informed Benjamin's understanding of the relation between knowledge and experience as infinitely continuous.[46]

In two fragments probably written in 1917 and 1918 as part of research for his dissertation, Benjamin contemplates the concept of the 'infinite task'. 'Science is in its form, (*not* its matter), an infinite task … [T]his does not mean a task with an infinite solution (temporally or otherwise). Infinite is the task [*Aufgabe*] that cannot be given [*gegeben*].'[47] Benjamin plays on the verb *geben* that is part of the word *Aufgabe*, literally, that which is given. The infinite task, he suggests, is, on the contrary, that which cannot be given. Science itself is the infinite task. Had it been 'given', or had it consisted of an answer to finite questions, it would have been contingent upon limitation and could not have consisted of an infinite task.

Science and knowledge conceived of as an infinite task are always open: 'every solution poses a new task.'[48] Benjamin uses Cohen's own conception of infinite task to criticize his understanding of science. While in 'On the Program' he redefined experience to include non-scientific experiences, here he redefines science against what he perceives as Cohen's empiricist approach.

In the second fragment, Benjamin criticizes Cohen's concept of infinite task for its being 'not a-priori' and 'empty'.[49] He describes the 'neo-Kantian' version of the conception of infinite task as a process the goals of which are determined over and over again in an infinite regression that empties the concept of meaning.[50] Benjamin's critique hinges upon the idea that an accumulation of a posteriori experience may not lead to *a priori*, absolute knowledge. The infinite task necessitates 'never-ending, absolute synthesis'.[51]

v The affinity of philosophy and teaching

As part of his critique of the Kantian and neo-Kantian's conception of knowledge as a product of rational reasoning alone, Benjamin puts forward a concept of teaching [*Lehre*] as encapsulating the continuity between experience and knowledge:

> But there is a unity of experience that can by no means be understood as a sum of experiences, to which the concept of knowledge [*Erkenntnisbegriff*] as teaching is *immediately* [*unmittelbar*] related, in its continuous development [*Entfaltung*]. The object and content of this teaching, this concrete totality of experience, is religion, which, however, is presented to philosophy in the first instance only as teaching. Yet the source of existence [*Dasein*] lies in the totality of experience, and only in teaching does philosophy encounter something absolute, as existence, and in so doing encounter that continuity in the nature of experience. The failing of the neo-Kantians can be suspected in its neglect of this continuity.[52]

Written several months subsequently to the main text, there is a shift of focus in the 'Addendum' towards a specific kind of experience, namely that of religion, which Benjamin describes as a totalized experience that serves as the source of existence [*Dasein*] – a term he had not previously used in the text. Benjamin's critique of the 'neo-Kantians' is not for lacking a conception of continuity – continuity is an important conception within Cohen's work – but for neglecting a certain type of continuity, that between knowledge and experience.[53] According to Benjamin's reading, the neo-Kantians presented knowledge as continuous within itself, delegating an inferior place to experience, which they viewed narrowly as scientific experience alone.

The italicizing of the word *immediately* suggests that by 'the continuity between knowledge and experience', Benjamin means their relation must be unmediated. Knowledge, in other words, consists of the natural development or unfolding as doctrine of this higher kind of experience. Benjamin views this continuity not only as epistemological but also as an ontological principle. Only experience that is continuous with knowledge may be totalized and become the source of existence. Since knowledge is ever changing, continually becoming, it necessitates experience. Existence – totalized experience – is described as the philosophical equivalent of religious teachings [*Lehre*], totalized knowledge.[54]

The concept of teaching is present in letters as well as essays Benjamin wrote to Scholem during this period. Thus, for example, in 1917 Benjamin writes:

> [I]t is in the philosophy of history that the specific affinity [*Verwandtschaft*] of a given philosophy with the true teaching should be able to come forth with maximum clarity, for this is where the subject of the historical evolution of knowledge for which doctrine is the catalyst will have to appear.[55]

Scholem describes Benjamin's use of *Lehre* during these years as related to the Hebrew 'torah' as instruction – 'instruction not only about the true condition and way of man in the world but also about the transcausal connection of things and their rootedness in God'.[56]

Benjamin contemplates these 'transcausal connections' or the 'affinity' between elements that are not linked in a causal relation in a fragment written in 1919 titled '*Analogie und Verwandtschaft*'.[57] In the text, Benjamin distinguishes between analogy, similarity and affinity. As an example for *Verwandtschaft* (which may also be translated as 'relationship' or 'kinship') he uses the affinity between parents and their offspring that is distinct from the similarities between them, nor is it based on the causal relation that connects them. Rather, 'affinity belongs, undivided, to the entire being without seeking for a specific expression [*Ausdruck*] (expressionlessness [*Ausdrucksloses*] of affinity)'. It is, in other words, an 'enigmatic' connection that links the internal being of elements. The concept of 'expressionless' [*Ausdruckslos*], through which Benjamin defines 'affinity', is further developed in his essay on Goethe's Elective Affinities. There, he defines the expressionless as a violently sublime moment that appears in the caesura or standstill of expression. 'Only the expressionless completes the work, by shattering it into a thing of shards, into a fragment of the true world, into a torso of the symbol', writes Benjamin.[58]

He differentiates *Verwandtschaft* from analogy, arguing that the latter is simply metaphoric similarity.

> The conflation of analogy and affinity is an utter perversion. It occurs either when analogy is regarded as the principle determining an affinity, or when an affinity is taken for the principle governing an analogy. We find the first type of confusion in people who imagine something while listening to music – a landscape, event, a poem. They seek something (rationally) analogous to the music … [I]t is impossible to move from music to an analogue; music recognises only affinity. Pure feeling [*reine Gefühl*] is related to music; this is knowable [*erkennbar*], and in it music is, too. The Pythagoreans tried to understand music by means of numbers.[59]

Confusion between affinity and analogy occurs for example when people attempt to express music in visual or verbal terms. Benjamin uses music to explain the difference between analogy and affinity – what may be expressed and the expressionless. It is not possible to move seamlessly between the visual and musical registers, just as it is impossible to move between analogy and affinity. Music, in its generation of affinities, is irreducible to numbers; it belongs to that type of non-scientific experience that Kant and the neo-Kantians did not recognize. This does not mean that music cannot be known, writes Benjamin. He distinguishes between two types of acquiring knowledge: through feeling and through ratio. While analogy is explainable by rational principles, affinity is 'knowable' through 'pure feeling'.

Through the concept of affinity Benjamin challenges Kant's conception of knowledge as a product of rational reasoning, suggesting sensation as a possible path to attaining knowledge [*erkennen*]. The affinity between two elements created by their shared connection to a third element seems close to Leibniz and Spinoza's conceptions of expression, in which modes and monads are linked through their participation in the being of God.

During this period, according to Scholem, Benjamin had attempted to formulate a 'symbols-doctrine' [*Symbollehre*] in which 'God is the unattainable centre of a system of symbols intended to remove him from everything concrete and everything symbolic as well.'[60] In fragments and notes from this period, the symbol is not defined as a triadic structure with direct relation or participation in the absolute; rather, the 'symbols-doctrine' consists of a hierarchic 'theory of orders'. Thus the 'highest order' includes a category of objects that 'fulfil themselves in objective intention' and 'point to God'.[61] Benjamin certainly does not call for a return to pre-Kantian speculation, and indeed, in another text from the same period, he critiques pre-Kantian philosophers for confusing experience and 'knowledge of experience'.[62] Although it is unclear to what extent the new conceptions of knowledge and experience Benjamin proposes in 'On the Program' and later works are speculative, there is a shift between the earlier fragments and the Foreword in his position vis-à-vis speculative experience.

Another fragment from the period explicitly defines the infinite task as an epistemological one par excellence. 'The true-being [*das* Wahr*sein*] (which as such is naturally unknowable) is linked with the infinite task [Auf*gabe*].'[63] The infinity of the task is here related to the unknowability of the true being it seeks. Its infinity lies in the impossibility of its completion. The tasks of epistemology are defined in the same fragment as the following: '1) The constitution of things in the "now of knowability" [*Jetzt der Erkennbarkeit*] and 2) the limitation [*Einschränkung*] of knowledge in the symbol'.[64] The 'now of knowability', a key term in the *Arcades Project* here used for the first time, highlights the significance of the 'now' – the passing – for experience and knowledge.

The symbol in which knowledge is limited is identified by Benjamin with Goethe's *Urphänomen*, according to which the essence of a phenomenon is perceptible in every one of its details.[65] This conception has a similar role to that of Leibniz's monad in his later work: both highlight the manner in which a totality of knowledge is condensed into symbols used in language. Symbol, *Urphänomen* and monad are all concepts that make manifest Benjamin's attempt to redefine knowledge as a process in which a totality is limited and abbreviated into details. However, as mentioned before, Benjamin also highlights the multiplicity of knowledge, and in this sense he diverges from Leibniz for whom the unity of knowledge derives from its source in God.

vi Cohen and Benjamin on origin, logical and historical

Through his conception of origin, Benjamin attempts to introduce a historical conception of singularity that is unrelated to genesis. He does so by describing origin as imperfect and incomplete; a continual, open-ended process. Benjamin does not refer directly to Leibniz's differential, yet his understanding of the singular is informed by Leibniz, mediated by Cohen's principle of origin, a methodological and ontological principle.[66] Benjamin retains the notion of origin as a key epistemological category, and as the productive force both of being and of itself, while rejecting Cohen's assumption that 'the productive force of thought can only manifest itself through the mediation of logic'.[67]

Leibniz's infinity as grounds for the finite serves as the nucleus of Cohen's origin. Cohen viewed infinitesimal calculus as the crux of scientific method, upon which the certainty of science depends. As he writes, '[t]he precise question and decisive answer for an indispensable, irreplaceable meaning of the thought as production must be obtained from the analysis of infinity.'[68] By positing infinitesimal calculus as origin, Cohen sought to overcome the foundation of Kant's epistemological doctrine in reality, leading to his distinctions between subject and object, sensibility and understanding. This was achieved through designing the principle of origin as inherent to the system of knowledge, which also serves as its ground. By recourse to infinitesimal calculus as the principle of becoming, Cohen overcame the strict separations present in Kant's 'Transcendental Idealism' between being and non-being, and between subject and object.

Cohen substitutes Kant's synthesis with origin, by replacing 'the question concerning the *connection* of A and B as given elements with the problem of the *origin* of B from that which is not'.[69] Nothingness thus becomes, in Cohen's thought, the very condition of being, just as rest, within the infinite system, contains an infinitesimally small movement or change: 'Pure thought produces the determinable itself, before its determination.'[70] Furthermore, Cohen viewed his principle of origin as distinct from the Romantics' logic of being, and from Kant's grounding of his doctrine, as Cohen understands it, on reality [*Realität*].[71] Thus instead of cognition being grounded in being, according to Cohen, being is grounded in thought, and 'thought is thought of origin'.[72]

Therefore, as Poma writes, on the one hand, the principle of origin is 'the supreme principle of pure knowledge'. All pure cognition must be a modification of the principle of origin, writes Cohen, since the logic of origin is the logic of pure cognition.[73] Yet, on the other hand, it is also a ground, or origin, not only of being but also of itself. Thought is thus productive of itself. In opposition to Kant and like Maimon, Cohen argues that nothing is given to thought; thought constitutes and grounds its own objects. The central instrument for this productive, constitutive thought, or logic of origin, is infinitesimal calculus.

Benjamin criticizes Cohen, in 'On the Program of the Coming Philosophy', for subordinating experience to thought and mathematics, and therefore favouring objective logical experience over subjective ephemeral experience. Yet like Cohen, Benjamin charges Kant with disrupting the continuity between experience and knowledge, between *a priori* reason and a posteriori experience. Benjamin's conception of origin consists of a return to Leibniz's infinitesimal, via Cohen's principle of origin, as the grounds for a conception of singularity based on a continuity between experience and knowledge.

Cohen highlights Leibniz's designation of thought as continuous, acknowledging the manner in which Leibniz applied his principle of continuity and reflections on infinity more generally to domains other than mathematics. To his mind, the significance of Leibniz's principle of continuity lies in its framing of relations between elements in nature and in thought through the infinitesimal method. The infinitesimal, in Cohen's view, is the fundamental concept of any truly modern logic of science, as he criticizes Kant for failing to perceive.[74]

Cohen credits Kant for having transformed metaphysics into 'critique' by applying the Newtonian system of principles to metaphysical thought.[75] And yet, he also challenges Kant's distinction between intuition and concepts, denying that intuition plays a role in knowledge acquisition and thereby rejecting Kant's understanding of a 'given reality' that exists prior to thought. Further, Cohen's conception of knowledge as dynamic and continually evolving was not compatible with Kant's schema of fixed *a priori* categories. This is one feature of Cohen's theory of knowledge which Benjamin takes up. Cohen's principle of origin embodies his conception of the evolving character (*Werdecharakter*) of science.[76] The principle of origin is a principle of thought, through which Cohen identifies thought with a process of becoming, grounding finite being in infinite thought. Thought does not consist, therefore, of representations or sensations; rather, it is a principle of the mathematical science of nature.

Origin functions, first, as an epistemological principle par excellence, since 'all pure knowledge cannot be but inflections of the principle of origin … [T]hus the logic of origin becomes that of pure knowledge'.[77] Second, origin serves as ground both of knowledge and of itself: thought is 'the productive activity itself …, the goal and the object of its activity'.[78] Thus there can be no 'givens', or Kantian 'things in themselves', since what produces being is thought itself. Benjamin too defines origin as becoming, in a reformulation of Cohen's principle of origin: 'By "origin" is meant not the coming-to-be of what has originated but rather what originates in the becoming and passing away.'[79] Benjamin describes the original phenomena as being defined by their incompleteness, just as every singular monad is an incomplete representation of the world from a different perspective. The dialectic inherent in origin shows that 'in everything essential, singularity and repetition prove to be reciprocally determined'.[80] Yet he denies that this conditioning of singularity upon repetition consists of a model of eternal return, as manifest in an early draft of the Foreword, where he defines a philosophical history that would concentrate on historical time:

> Such a conception of history, for which the border between natural and world history should become severely problematic, would regard repetition as an essential moment in every form of periodisation that occurs in natural and world history, and would turn the question in what sense repetition can appear in history – unrepeatable in itself – into the *experimentum crucis* of its philosophy of history.[81]

While the question of the form of repetition is a crucial one for such a philosophical history, history in itself, as a whole, is unrepeatable.

'The category of origin is not therefore, as Cohen holds, a purely logical one, but a historical one', writes Benjamin.[82] He locates the origin not in thought, as does Cohen, but rather in the 'fore and after history' of the idea. That is, in the virtual past and future contained by the idea as the monad is 'pregnant' with its future.[83] Benjamin's concept of origin, therefore, returns, in a sense, via Cohen to Leibniz. Origin becomes a category not of thought itself but of the history of thought and science, and thus of history more generally, corresponding with Leibniz's emphasis on the historical nature of phenomena, which contain both marks of their past and a 'script' detailing their future.[84]

2 Minute perceptions

i Differentials of consciousness

Deleuze turns, as well, to Leibniz's differential calculus in an attempt to reformulate a category of singularity grounded in a continuity between experience and learning. His return is informed by Maimon, who, like Cohen, returned to Leibniz as a way to contest Kant's separation between the process of cognition and the 'thing in itself'. Challenging the traditional conception of knowledge as something that can be possessed, Deleuze replaces the notion of knowledge acquisition with a continual process of learning in which there are no hierarchies between the learner, the learned and the process of learning.

In his analysis of Leibniz in *The Fold*, Deleuze highlights, as Benjamin does implicitly, the permanent state of incomplete perception of every monad.[85] Leibniz's theory of 'minute perceptions' serves as a focal point in Deleuze's reading of Leibniz therein, as well as in *Difference and Repetition*. By focusing on Leibniz's minute perceptions, Deleuze suggests 'in between' states of consciousness like sleep, fainting or hallucination as models for learning. As Deleuze writes elsewhere, '"learning" always takes place in and through the unconscious, thereby establishing the bond of a profound complicity between nature and mind'.[86] He highlights, following Maimon and Leibniz, differential calculus as a process by which thought is generated internally. For Deleuze, singularity is a category of thought as well as an ontological category, not one of history as in Benjamin. In this sense he comes closer to Cohen's 'logical interpretation' of Leibniz.

Grounded in the logic of infinitesimals guiding his calculus and dynamics, Leibniz put forward a revolutionary theory of unconscious, known as the theory of 'minute perceptions'. Just as bodies, according to Leibniz, are never completely still, rather, when seemingly still they are in fact minutely dynamic, so perception never ceases from its activity. 'Minute perceptions' consist of a constant flow of impressions made on us by exterior bodies that in some cases enter our conscious realm. As Leibniz writes,

> These minute perceptions, then, are more effective in their results than has been recognized. They constitute that *je ne sais quoi*, those flavours, those images of sensible qualities, vivid in the aggregate but confused as to the parts; those impressions which are made on us by the bodies around us and which involve the infinite; that connection that each being has with all the rest of the universe.[87]

Minute perceptions consist of the scattering of seemingly negligible impressions through which other bodies affect our own, expressing the continuity between our consciousness and the universe. They vary in degrees of intensity, becoming more vivid when coming together and entering our consciousness. Leibniz often uses the image of the sound of roaring waves to illustrate these ideas. When standing on the beach we hear the sound of many waves merged together unconsciously; it enters our consciousness only when a single wave touches us or affects us in a specific way. The fact that we are not conscious of our initial perception of the subdued sound of multiple waves does not mean this sound is not perceived, minutely, or with diminished intensity.

Deleuze interprets Leibniz's theory of minute perceptions as describing the relations between the remarkable and the ordinary. He understands the differential relation between minute perceptions and conscious perceptions as one between the remarkable and the regular, rather than part-whole relations. The wave we become conscious of is the remarkable, roaring one, not the regular ones whose sounds blur into each other. 'We must understand literally – that is mathematically – that a conscious perception is produced when at least two heterogenous parts enter into a differential relation that determines a singularity.'[88] Singularity is produced, then, by the understanding of differential relations as active internally, determining the process of moving from unconscious to conscious experience.

'Every consciousness consists of a threshold', writes Deleuze in *The Fold*. The threshold is crossed when 'two heterogenous elements at least enter into a differential relation that determines a singularity'.[89] In order to discern singular elements within the flow or confused perceptions, two or more perceptions must interact in a way that would produce a third, conscious perception. The relation between the perceptions is differential; that is, the elements are determined only by their relation to one another, and this determination is internal to their reciprocal relation.[90] Instead of a conscious perception being related to an object in space, as per Kant, it is related to a multiplicity of minute unconscious perceptions as a singular relates to the ordinary or unremarkable. The conscious perception is thus determined by the differential relation between minute unconscious perceptions.

Deleuze finds inspiration in the aforementioned Jewish philosopher Salomon Maimon, a contemporary of Kant's who critiqued Kant's philosophy and earned his appreciation.[91] As part of his 'sceptical rationalism', Maimon 'uses Leibnizian elements in order to amend critical philosophy by eliminating its weaknesses'.[92] In other words, Maimon adopts some elements from Leibniz (or 'the Leibniz-Wollfian system' in his words) in order to overcome Kant's opposition between understanding and sensibility, concept and intuition. As such, his theory satisfies Deleuze's requirement for enabling an immanent model of learning.[93]

Maimon attempts to solve the noted paradox of Kant's 'thing in itself', namely the distinction he makes between noumena and phenomena. Kant is famously criticized for the circularity of his argument, which assumes *a priori* facts of knowledge such as mathematical truths, and then seeks their condition of possibility. In assuming *a priori* knowledge that serves as the foundation of knowledge of the transcendental subject, Kant posits a condition of knowledge that is external to reason itself. As Maimon puts it,

> How can the understanding subject something (the given object) to its power (to its rules) that is not in its power? In the Kantian system, namely where sensibility and understanding are totally different sources of our cognition, this question is insoluble as I have shown; on the other hand, in the Leibniz-Wollfian system, both flow from one and the same cognitive source.[94]

Maimon replaces Kant's opposition between understanding and sensibility with levels, or degrees, within consciousness, thus offering an immanent understanding of epistemological processes.[95] He does so by putting forward a concept of 'differentials of consciousness', inspired by Leibniz's differential calculus and concept of 'minute perceptions'.[96] The differential, which measures the rate of change of a function for example, allows Maimon to replace Kant's binary with an internal cognitive scale. But while Leibniz uses the differential to describe the relation between conscious and unconscious perceptions, Maimon describes the relationship between *noumenon* and *phenomenon*, perception and object, or thing, as differential.[97]

As an example, he writes, the representation of the colour red should be thought of not as a finite extension but as the 'differential of an extension'.[98] Moreover, the word 'representation' itself is erroneous – it should be considered as the 'presentation' of what had not existed before, not the 'representation' of what had already existed.[99] The perception of red is generated differentially, or genetically, as Maimon also puts it, resulting in a new cognition of the red object.

As Deleuze describes it in *The Fold*, for both Leibniz and Maimon, differential determination does not refer to divine consciousness but rather to minute perception as expressing the infinite within the finite self. This dynamic, or tension between the infinite and the finite, Deleuze writes, is 'precisely the Baroque position of equilibrium, or disequilibrium'.[100] The immanence of infinite knowledge within the self is an elegant solution that, however, creates uneasy tensions for Leibniz and Maimon. Deleuze seeks, as mentioned, to overcome the limitation of the infinite by the finite and proposes a limitless infinity.

Although Deleuze's references to Maimon in *Difference and Repetition* and *The Fold* are brief, they show how significant Maimon was for Deleuze's readings of Leibniz. Maimon and, with him, Deleuze argue that for Kant, difference is external to both determinable intuition and determining concept. According to Maimon, *a priori* knowledge must be itself engendered by reason. Deleuze conceives this very determination as a dynamic process that takes place through what he calls 'degrees of intensity'.

ii Degrees of intensity

Deleuze describes the internal concept of difference, which is determined by differential relations, as 'intensity'. As he writes, 'intensity is the form of difference in so far as this is the reason of the sensible. Every intensity is differential, by itself a difference.'[101] Deleuze's concept of intensity is shaped, once again, by 'correcting' a Kantian concept, namely that of 'intensive qualities', by recourse to Leibniz.

In the section on 'Anticipations of Perception' in the First Critique, Kant defines 'intensive qualities' in the following: 'In all appearances, the real, which is an object of the sensation, has intensive magnitude.'[102] That is, objects of experience have degrees of reality that are determined independently from their extension in space. This is the only place where Kant explicitly refers to intensive magnitude as 'flowing' [*fliessend*].[103] This 'dynamic' understanding of the objects of experience views matter as irreducible to size and shape alone.[104] Deleuze's concept of intensity is based on this dynamic understanding of degrees of reality, and he criticizes Kant for designating 'intensive quantity' as only applying to matter that fills a given extensity.[105]

Leibniz's understanding of intensity, as the varying degrees of sensible qualities that compose perception, fulfils the demands of the concept of intensity sought by Deleuze, one that is independent from extension. According to Leibniz, perceptions are actualized, becoming conscious and therefore more intense, through differential relations. Their degrees of intensity are therefore determined by difference, which corresponds with Deleuze's model of intensity as 'difference in itself'.

In a famous critique of Leibniz's theory of absolute space, Kant described the difference between two enantiomorphs, such as that between the left and right glove of the same person, as internal.[106] Yet, as Deleuze notes, since this difference is non-intrinsic and non-conceptual, it cannot truly be internal – that is intensive. As it relates to extension in space, it can only be represented in the opposition between each hand and extensity as a whole.[107]

Deleuze argues rather that such difference between two enantiomorphs, the 'paradox of symmetrical objects', must have an intensive source, citing Cohen as correctly seeking to reformulate 'intensive quantities' in his interpretation of Kant.[108] In Leibniz, on the contrary, as Deleuze notes, there is an affinity of extrinsic differences and intrinsic, conceptual ones, since all predicates are contained in the concept, which in effect renders all differences conceptual. This affinity is grounded on an internal, intensive, differential process.[109]

iii Learning to swim

Using Leibniz's conception of 'minute perceptions', Deleuze describes the idea as a system of relations between differentials, and learning as an interaction between our minds and bodies, on the one hand, and singular points within the 'problematic idea', on the other:

> The idea of the sea, for example, as Leibniz showed, is a system of liaisons or differential relations between particulars and singularities … [T]o learn to swim is to conjugate the distinctive points of our bodies with the singular points of the objective Idea in order to form a problematic field. This conjugation determines for us a threshold of consciousness at which our real acts are adjusted to our perceptions of the real relations of the object, thereby providing a solution to the problem.[110]

Deleuze designates the point of shifting from unconscious into conscious perceptions, in which the singular is engendered by differential relations (named in *The Fold* 'the point of inflection'), as the point at which the problem is solved, and something is learnt. He describes relations of expression between our conscious and unconscious perceptions, our actions, and the 'real relations of the object'. Deleuze's 'Problematic ideas' consist of elements of nature and at the same time comprise particles of our unconscious perceptions, marking a 'profound complicity' between nature and soul.

While Benjamin describes the inner continuity between experience and knowledge as mirrored by an external continuity between history and nature, for Deleuze there is no strict separation between internal and external, consciousness being connected in a non-mediated manner to nature. Leibniz's theory of minute perceptions is significant for Deleuze's critique of Sigmund Freud's unconscious and his own conception of the unconscious developed in *Anti-Oedipus* and *A Thousand Plateaus*. In a lecture, he credits Leibniz with offering one of the first theories of unconscious and suggests that Freud drew on a long tradition of theorizing the unconscious that originated in Leibniz, and that the novelty of Freud's theory of unconscious was in its creating an opposition between consciousness and unconsciousness instead of Leibniz's differential relation between the two. Carl Jung, Deleuze says, reintroduced a notion of differential consciousness, to Freud's discontent.[111]

In *Difference and Repetition*, Deleuze opposes the 'image of thought' to other images such as a schizophrenic image, suggesting schizophrenia as an undogmatic way of approaching the world, a 'possibility for thought – one, moreover which can only be revealed as such through the abolition of the image'.[112] As we will discuss in Chapter 4, Deleuze links repeatedly between schizophrenia and Mannerism, describing the former as the trait of the latter. This implies that the Baroque-Mannerist 'image of thought' is in a sense opposed to the one constructed by Western philosophy. Baroque is viewed, that is, as a uniquely heretical moment in the history of philosophy and science, rather than as the period in which the foundations for the philosophy of the Enlightenment were laid.

iv Individuation, expression, crystallization

Deleuze differentiates between two interpretations of Leibniz: 'Cartesian Leibniz', on the one hand, and 'Nietzschean Leibniz', on the other. He asks: How might we understand Leibniz's statement that a clear idea is clear inasmuch as it is confused? The Cartesian Leibniz would understand this as meaning that the clear idea is confused because it is not clear enough. Leibniz in this case follows Descartes' understanding of human ideas as extending across a range between the clear and the confused.

The Nietzschean Leibniz would rather suggest that there is a difference of nature, not of degree, between clear and distinct, hence the clear is in itself confused, and the distinct in itself obscure. An idea that is clear inasmuch as it is confused, or, as Deleuze describes it, 'the properly philosophical stupor of the Dionysian idea' is another way of describing Deleuze's 'Problematic Idea'.[113]

In 'Asymmetrical Synthesis of the Sensible', the closing chapter of *Difference and Repetition*, it seems Deleuze opts for the second, 'Nietzschean' interpretation of Leibniz. This interpretation then plays an important role in Deleuze's forming of his own theory of individuation.

> Only Leibniz approached the conditions of a logic of thought, precisely inspired by his theory of individuation and expression. Despite the complexity and ambiguity of his texts, it does indeed seem at times that the expressed (the continuum of differential relations or the unconscious virtual Idea) should be in itself distinct and obscure: for example all the drops of water in the sea, as genetic elements and their differential relations, the variations of these relations

> and the distinctive points they comprise. And that the expressor (the perceiving, imagining or thinking individual) is by nature clear and confused: for example our perception of the noise of the sea, which confusedly includes the whole, yet clearly expresses only certain relations and points by virtue of our bodies and the threshold of consciousness they determine.[114]

Instead of the idea being clear and distinct, that which is expressed is both distinct and obscure, and the expresser is clear and confused. The relation between the genetic elements, or the noise of drops of water, and the sound of the sea as a whole is not one of part-whole but rather of virtual-actual. Once the tiny perceptions of the sounds of drops cross the threshold of consciousness, they become actual. Thus, human experience in general, as well as the process of the posing and solving of Problematic Ideas, is best described, according to Deleuze, as one of actualization rather than of unification or synthesis.

Although Deleuze criticizes Leibniz's inclinations towards identity and perfection, it is clear that in Leibniz's metaphysics and epistemology he finds an alternative to Kant's transcendental subject. Deleuze's conception of individuation, alongside its source in Gilbert Simondon's theory of individuation, is informed by Leibniz's conception of the individual substance, which is constituted by its differential relations to other individuals that make the system of the world.[115] Individuation and differentiation are two discernible yet interlinked processes, according to Deleuze:

> Individuation does not presuppose differentiation, it gives rise to it. Qualities and extensities, forms and matters, species and parts are not primary, they are imprisoned in individuals as in crystals. And it is the world itself, that may be read as in a crystal ball in the moving depth of individuating differences or differences of intensity.[116]

The constitution of the individual and that of the world are interdependent, as in Leibniz's expressive model. The individual includes its qualities and parts 'as in a crystal', while the world and its events may be read within the fluctuations of intensity that constitute the individual. Dosse and Buci-Glucksmann trace the omnipresence of the crystal in Deleuze's work, the crystalline plane serving as a model of his concept of the 'event', as Dosse writes, to his reading of Leibniz.[117] Indeed, in the above passage, the crystal embodies Deleuze's version of Leibniz's 'living mirror' metaphor, emphasizing the capacity of a crystal ball to reflect events future and past.

Deleuze draws on Leibniz, yet, in the same breath, distances himself from him by defining difference via the terms 'expression', 'implication' and 'envelopment', used when discussing Leibniz's theory of expression in *Expressionism in Philosophy* and later in *The Fold*. Thus, in *The Fold*, implication is described as a key characteristic of the individual monad: the monad implicates the world that does not exist outside of the monads that express it. In the final chapter of *Difference*, intensities function both as 'enveloped' and as 'enveloping'; they are folded into one another, so to speak. As enveloping, intensity expresses the world clearly, while when functioning as enveloped, it expresses the world confusedly.[118] The intensities that are enveloping belong to the individual, or thinker, and are *both* clear and confused; those that are enveloped belong to the idea, and are *both* distinct and obscure.

Thus Deleuze uses Leibniz's conception of expression, in its Nietzschean interpretation, to transcend the binary thinking that, he claims, has plagued the philosophy of representation since Plato. While in Descartes, clear-and-distinct describes the mind, and confused-and-obscure describes the body, in Leibniz, according to Deleuze's reading, clear-and-confused relates to mind and distinct-and-obscure to body. The location of the monad on the range between clear and confused perception is defined as its perspective, that which marks it as a unique individual. Deleuze uses these aspects of Leibniz's metaphysics in order to offer a model that forms a continuity between mind and body.

Deleuze concludes the chapter, and *Difference and Repetition* as a whole, with a brief statement in which the interaction between the individual and his or her other are defined through the conceptions of expression and implication. Instead of designating the other [*L'autrui*] as either object or subject, Deleuze argues, the other is 'defined in each system by its expressive value – in other words, by its implicit and enveloping value'. The 'other' is defined by the way in which it expresses a possible world, that is, by how it is implicated in this world and this world is implicated in it.

The multiplicity of possible worlds are in themselves our 'others', and hence, in order not to either objectify or subjectify others, we must consider them as the expression of possible worlds. Language is described by the means by which the other makes such possible worlds real, and as such, it is the source of lies. On the other hand, language itself functions as an expressive other, or an enveloping centre. The more expressive it is, the higher its tendency to internalize difference.[119]

3 Learning as recollection; Plato, Benjamin Deleuze

Benjamin and Deleuze take issue with Kant's theory of experience and, inspired by other critics of Kant, return to Leibniz's pre-Kantian theory of knowledge and infinitesimal calculus in order to rectify Kant's critical philosophy, and dispense with its binary formulations. Both the Foreword to Benjamin's *Trauerspiel* book and Deleuze's third chapter of *Difference and Repetition*, 'The Image of Thought', are profoundly informed by the idealist tradition, specifically Leibniz and Plato. At the same time, they strongly reject this tradition in order to offer an alternative philosophy. For Benjamin, this is a philosophy of history tasked with the analysis of works of art, while for Deleuze, this is a philosophy of difference.

These distinct aspects of their theories of learning are manifest in their respective references to the archetypal theory of learning presented in Plato's *Meno* and discussed by Leibniz in his *Discourse on Metaphysics*.[120] The dialogue is renowned for describing learning as a process of recollection. Socrates suggests to his opponent, Meno, that learning consists of the recollection of knowledge that the soul innately possesses from its previous life. The knowledge in question is mathematical knowledge, which often serves as an example for the archetype of absolute knowledge.

In the well-known passage from the dialogue, Socrates illustrates his argument by drawing a square with a stick in the sand and asking a slave boy a series of questions in geometry in relation to its dimensions. The boy replies correctly to his geometrical queries, demonstrating, according to Socrates, that the process of learning is 'nothing but recollection' activated through questioning, rather than teaching.[121] He also justifies his own method of asking his opponents questions until they reach an *aporia*, or state of puzzlement or doubt.

Benjamin compares the concern of philosophy, described as the restoration of 'the symbolic character of the word' through the recollection of 'primordial perception' [*Urvernehmen*] to Platonic *anamnesis*. 'Only it is not a matter of an intuitive making-present of images; in philosophical contemplation the idea as word – as the word that claims anew its naming rights – is released from the innermost reality.'[122] For Benjamin, philosophy is tasked with restoring a specific kind of knowledge: that of the symbolic, non-communicable aspect of words. He refers to Plato's dialogue in order to highlight that philosophy is a kind of learning that takes place in time, through a process of recollection aided not by visual aids as in *Meno* but through words. Through the representation [*Darstellung*] of words in philosophy they regain their symbolic, primary character in which 'the idea comes to a self-understanding'.[123] This process of recollection is also compared to the act of naming, in which knowledge is recovered through an affinity between the name and what is named. In the earlier text 'On Language as Such and the Language of Man', Benjamin described the act of naming as participation in divine knowledge. As he wrote, 'The deepest images of this divine word and the point where human language participates most intimately in the divine infinity of the pure word, the point at which it cannot become finite word and knowledge, are the human name'.[124] In this text, which we cannot fully analyse here, Benjamin marks the origins of language in *Genesis* and the fall from Eden. Man lost the immediacy of communication [*Mitteilung*] of the name in the fall, tumbling into 'the abyss of prattle', as Benjamin writes.[125] It is only after the fall that language became mediate, writes Benjamin, paving the way for the tower of Babel.

By evoking *Meno* in the Foreword, Benjamin alludes once more to the idealistic or speculative aspect of linguistic knowledge that lies in opposition to external communication. He critiques the narrow conception of knowledge as mathematical knowledge that can be represented visually. Philosophical truths, he argues, are revealed in this non-communicable aspect of language rather than through mathematical deduction. Benjamin reformulates, in the Foreword, the conception of the 'idea', which, following Plato, he views as foundational for philosophical work, through a critique of Plato's *eidos* or visual form, suggesting that ideas function on a non-visual, non-communicable level.

In *Logic of Sense* and *Difference and Repetition*, Deleuze defines the very task of modern philosophy as the reversing of Platonism.[126] Yet, while he views Plato as one of the founders of the philosophy of representation, Deleuze does not dismiss every aspect of his philosophy. In the chapter 'The Image of Thought' he alludes to *Meno* as part of his critique of the traditional conception of knowledge. As mentioned, Deleuze distinguishes between knowledge attained by apprenticeship on the one hand and learning on the other.

While knowledge is gained through rudimentary empirical experience, using the principle of identity, learning is a transcendental process produced by and creating differences. In *Meno*, Deleuze writes, there is a true learning process of the soul, and yet,

> a new Meno would say: it is knowledge [*le savoir*] that is nothing more than an empirical figure, a simple result which continuously falls back into experience; whereas learning [*l'apprendre*] is the true transcendental structure which unites difference to difference, dissimilarity to dissimilarity, without mediating between them; and introduces time into thought – not in the form of a mythical past or former present, but in the pure form of an empty time in general.[127]

Deleuze describes learning as a structure of relations in which differences are brought to bear upon one another. He follows Plato in depicting learning as a movement of the soul in time (in its pure form); hence learning cannot be reduced to knowledge. Like Benjamin, Deleuze describes learning as immediate, and as a passage through the unconscious. At the same time, Deleuze criticizes Plato's understanding of knowledge as innate and absolute, while in his symbolic theory of naming, Benjamin approaches a notion of absolute knowledge.

Leibniz describes Plato's doctrine of reminiscence as 'very solid'.[128] In his own interpretation of *Meno*, Leibniz reiterates his theory of expression: 'Our soul expresses God and the universe, and all the essences as well as all existences ... nothing can be taught to us whose idea we do not already have in our mind, an idea which is like the matter of which that thought is formed.'[129] Plato's doctrine of reminiscence is true, Leibniz suggests, as long as we understand it correctly. Our notions come from internal ideas rather than external experiences; it pre-exists in our minds as a confused, unmoulded matter. Learning or 'remembrance' of knowledge is equivalent to moving from confused to distinct ideas.

Deleuze and Benjamin's theories of learning challenge any such conception of innate knowledge. While both draw on Leibniz's infinite calculus, as we have seen, Benjamin's concept of expression is based on an interpretation of Leibniz's monad as a limitation of the infinite, and Deleuze calls for an unlimited expression. Leibniz's concept of the virtual, which derives from his doctrine of innate ideas, is, however, of interest for Deleuze and will be further explored in the next chapter.

3

Spectres of eternal return: Baroque forces in Benjamin's theory of historical experience

Introduction: Arcades monadology

'Whenever a dialectical process takes place we are dealing with a monad', writes Benjamin in 'Convolute N', entitled 'On the Theory of Knowledge, Theory of Progress' of the *Arcades Project*.[1] This 'convolute', or section, of the monumental *Arcades Project*, upon which Benjamin worked during the last decade of his life, contains material that may have been in part intended for inclusion in what Benjamin describes as an epistemological introduction to the work comparable to the Foreword to *Origin of the German Trauerspiel*.[2] As Benjamin writes in another fragment in the convolute, since the objects of history are structured as monads, the dialectical process of historiography engages with monadic structures.[3] This chapter interrogates the role of monadic structures in this and other texts by Benjamin that concern questions of historical experience. These allusions to Leibniz's philosophy, we will see, uncover tensions between repetition and exceptionality that are often conveyed through music-related metaphors and images.

Benjamin's use of Leibniz's term 'monad', here and in *On the Concept of History*, is surprising for two reasons. First, because of the way it is conjoined with the term 'dialectics': if there were a clear-cut division between Benjamin's earlier writing, often defined as 'metaphysical' or 'idealistic', and his later historical-materialist concerns – such formulations would not exist.[4] Furthermore, in his late philosophy of history, written in the 1930s, Benjamin insists on the 'discontinuity' of history, which the term 'dialectics' implies as well. The monad, some would suggest, primarily denotes continuity between the expresser and the expressed. The second reason it is surprising is because we can assume several years have passed since Benjamin's engagement with Leibniz's philosophy. As noted earlier, there is some evidence that Benjamin read Leibniz, as well as commentaries on Leibniz, as a student, while there is no verification of his having read Leibniz after completing *Origin of the German Trauerspiel* in 1925.

This perhaps unexpected use of Leibniz's conceptions in Benjamin's later writing manifests the manner in which Leibniz's monad, as an expressive, intensive substance endowed with force, links Benjamin's earlier and later philosophy of historical knowledge and experience. In this vein, the present chapter suggests that Leibniz's conception of force, specifically what he calls 'living force', and the internal dynamic within the monad between 'virtuality' and 'actualization', underlie Benjamin's theory of historical experience. Benjamin, like Deleuze, is drawn to Leibniz's structure of expression as the spatial equivalent of eternal return: the notions of a multiplicity of worlds simultaneously existing within our own (Deleuze) or an infinity of ideas, each containing an image of the world (Benjamin), replace the infinite repetition of this world. This structure opens up space for virtual, other worlds that do not exist in actuality but enable heterogeneity within our world.

Based on his principle of conservation of force, the conception of force forms part of Leibniz's understanding of substance.[5] Every monad contains a self-generated force that drives it to action. Motivated by the monad's 'appetition' or conatus, this force carries it from one perception to the next. Force determines, then, both the manner in which the monad acts and the manner in which it perceives or knows the world (which coalesce to some extent in Leibniz's philosophy). Tending towards perfection, active force drives the monad towards ever more clear and distinct perceptions.[6] In itself, force is never completely actualized; rather, it continuously tends towards actualization. Driven by this force, the substance 'unfolds' moving from virtual to actual knowledge, and from virtual or unconscious perceptions to actual, conscious ones.

This movement of actualization motivated by force informs Benjamin's philosophy of history, and specifically his concepts of the dialectical image and history as an image-montage. Force functions, then, both on an inter-monadic level – within the image-monad – and on an intra-monadic level – determining the relations between multiple image-monads, as it does in Leibniz's philosophy. In this manner, the 'actualisation' of history Benjamin describes in the *Arcades Project* may be understood as relying on an understanding of history past and future as virtually contained within the dialectical image. Leibniz's monadic force and questions of continuity and discontinuity, singularity and repetition also underlie Benjamin's concept of violence [*Gewalt*] and Baroque theory of sovereignty. These are outlined in response to Carl Schmitt's theory or sovereignty, as expounded in *Political Theology*.[7] Unlike in Leibniz, where force is infinite, in Benjamin's analysis of Baroque plays, immanent force that has no outlet or limit becomes violent.

The final sections of the chapter will interrogate how, as part of his theory of historical experience, Benjamin engages with the threat of determinism that accompanies the appropriation of Leibniz's conceptions of expression and force. These concerns appear in Benjamin's interpretation of Mallarmé's *Un coup de dés jamais n'abolira le hasard*, and his recurrent references to the notions of 'apokatastasis' and 'eternal return'. In the *Arcades Project* and his essay on Leskov, 'apokatastasis' is presented as a form of redemption through a shift in vision – a restoration rather than repetition of the same.[8] On the other hand, Benjamin criticizes, in the *Arcades Project*, Louis Auguste Blanqui's versions of eternal return. While, like Benjamin, Deleuze alludes to Mallarmé and Nietzsche in relation to Leibniz, his concept of 'eternal return' sets him apart from Benjamin. For Deleuze, Nietzsche's eternal return, in which chance and necessity are inseparable, presents a way beyond Leibniz's notions of necessity and contingency.

1 Baroque historian

i Perfect history

Leibniz composed a large and varied output of historical writings during his lifetime, yet, often considered less innovative than his scientific and philosophical ones, his work as a historian remains understudied.[9] Arguably, while Leibniz's historical texts had limited influence on historiography, his philosophical writings were imperative for the theorization of history as a unified system expressed in epic form by eighteenth-century German Enlightenment thinkers.[10] Leibniz's direct references to historical method are not elaborated comprehensively. As in many cases with Leibniz, since there are contradictory statements in his writing, different standpoints have been emphasized by critics, which has led to highly divergent interpretations of his thinking on history. Some of his earlier readers extracted a philosophy of history based on Leibniz's principle of continuity from his work, arguing that Leibniz sought to unearth a continuity between past and present events through which the prediction of future events would be possible.[11] A view of history as structured according to the principle of continuity would imply that humankind as a whole continually progresses towards perfection. Yet this is an issue over which Leibniz had apparently changed his mind, which renders the interpretations of Leibniz's view as necessarily based on continuity and progress towards perfection questionable. Between 1693 and 1715, Leibniz composed a series of texts in which these questions were investigated.[12] While in 'Concerning the Horizon of Human Knowledge' (1693) he argues for eternal return in a manner contradicting the idea of humankind's constant progression, in a text from 1715 he seems once again committed to the idea of human progress.[13]

In *Theodicy* (1710), his best-known work during his lifetime, Leibniz viewed nature and history as originating in God, and therefore history as fulfilling the divine plan.[14] Regardless of whether Leibniz changed his mind about the progress of humankind, his conception of substance as constantly developing towards perfection in an infinite progress, in a manner that is in a sense predetermined, informed Kant's well-known essay 'Idea for a Universal History with a Cosmopolitan Purpose', and consequently had a significant (albeit second-hand) influence on the Historicists.[15]

As Koselleck argues, Leibniz was the first to temporalize the process of becoming complete or perfect and render it into a worldly series of occurrences brought about by humans, with objectives that are transferred from one generation to the next.[16] Before Leibniz, perfection was viewed as attainable only in the next world. Leibniz's understanding of the individual as endowed with active force, its freedom consisting in a degree of spontaneity, underpins his view of the moral perfection of humankind as constantly developing. These informed, *inter alia*, the concepts of individualism and progress developed by Kant, Herder and the Historicists in their philosophies of history.[17]

Leibniz described the goal of both history and poetry, in the *Theodicy*, as moral: more precisely, it was to 'teach wisdom and virtue by example': the history of humanity as God's novel. Koselleck views this as an example of the coalescence of the demands of history and poetry, following a process which saw historical narrative as providing the unity of the epic, as evident in the way the plural form *die Geschichte* was condensed into a collective singular denoting history as a system.[18] Leibniz's metaphor of the novel was taken up by Kant, who uses the term *Roman* to express the desired systematic unity in the seemingly random aggregate of human actions.[19]

The *Monadology* was in all probability Leibniz's work with which Benjamin most thoroughly engaged.[20] Coudert argues convincingly that the Lurianic Kabbalah influenced Leibniz during his later years, an influence that affected most of the ideas presented in the *Monadology*.[21] It seems probable that

Benjamin, who himself had an interest in Kabbalah and other gnostic sources, was drawn to these elements in Leibniz's *Monadology*.[22]

The coalescence of poetry and history, both described as 'epic forms', are recurrent themes in Benjamin, despite his opposition to Kant and the Historicists' teleological account of history as system. As we will see, in its particular ability to make use of images, or its imageability [*Bildlichkeit*], poetry is viewed by Benjamin as a predecessor of cinematic montage, which poses a challenge to the narrative form of the novel. Thus although Leibniz's conceptions of progress and individuality – and the manner in which they were developed by Kant and the Historicists – were rejected by Benjamin, aspects of Leibniz's late metaphysics and dynamics were significant for Benjamin's critical thinking on history, and the way in which it was articulated through concepts of the image.

ii Living force

As noted, force is of paramount importance in Leibniz's understanding of the simple substance, or monad. As Leibniz writes: 'the concept of forces or powers, which the Germans call *Kraft* and the French *la force*, and for whose explanation I have set up a distinct science of *dynamics*, brings the strongest light to bear upon understanding of the true concept of *substance*.'[23] In the *Monadology*, Leibniz describes monads as 'entelechies' or 'perfection-havers', an Aristotelian term Leibniz appropriates and reformulates: while for Aristotle it denotes a realized *state* of perfection, Leibniz uses the term to describe a substance containing within it the principle of its own change – a germ of perfection which it tends to develop.[24] In other words, one of the principle characterizations of Leibniz's simple substance or monad is its force or tendency towards perfection.

Leibniz refers to force from his early works and on, and it receives full elaboration in his late essay 'Specimen Dynamicum' from 1695.[25] Contra Descartes' contention that extension is the only essential attribute of bodies, Leibniz argues that prior to extension substances encompass force. Leibniz believed that explanations of natural phenomena must include, beyond the empirical laws of motion, dynamics that explain these laws and are founded on metaphysical truths.

He distinguishes between two kinds of force; the first, primitive force, is that of suffering or resisting. The second, active force, is described by Leibniz in the following way:

> Active force … contains a certain act or entelechy and is thus midway between the faculty of acting and the act itself and involves a conatus. It is thus carried

> into action by itself and needs no help but the removal of an impediment. This can be illustrated by an example of … a bent bow. For though gravity and elasticity can and ought to be described mechanically by the motion of ether, the ultimate reason for motion in matter is nevertheless the force impressed upon it in creation, which inheres in every body but is variously limited and restrained in nature through the impact of bodies upon each other.[26]

'Active force' may be described, then, as a kind of potentiation of action that constantly tends towards realization. It originates in God, who imprints substances with force upon their creation, yet it also includes an element of self-action of the body in question and can act upon other bodies when unimpeded. Inherent in all substances, active force produces a minute level of action at all times, so that bodies are never in complete rest. Leibniz further distinguishes between 'dead' and 'live' active force; while the former is purely mechanical force, such as centrifugal force or gravity, the latter, *vis viva* or living force, is grounded in Leibniz's infinitesimal system and described as an amassment of an infinite number of impressions of dead force. Live force is therefore infinitely graded or toned; it is a potential that is continuously actualized or *unfolded* to infinity.[27]

Leibniz's theory includes an unresolved tension by his designation of force, on the one hand, as self-movement, and, on the other, as 'imprinted' in substances by God and controlled by a pre-established harmony. As he writes in 'Principles of Nature and Grace' (1714):

> For all is regulated in things, once and for all, with as much order and mutual connection as possible, since supreme goodness and wisdom can act only with the perfect harmony. The present is big with the future, the future might be read in the past, the distant is expressed in the near. We might get to know the beauty of the universe in each soul if we could unfold all that is enfolded in it and that is perceptibly developed only through time.[28]

Leibniz describes the world as ordered in the best possible way thanks to God's endowing it with a 'perfect harmony'. The present contains the future enfolded within it, yet we can only perceive its unfolding with time. The self-movement of bodies consists uniquely, then, of their moving from one perception to the next towards a more perfect representation of the world, or, in other words, of their executing the plan [*détail*] imprinted upon them by God.

iii The virtual

An important aspect of Leibniz's theory of force is the category of the 'virtual', which Leibniz opposes to the 'active' or 'actual', yet rather than an Aristotelian potency or a capacity, it signifies something like a tendency towards action.[29] In the introduction to the *New Essays on Human Understanding*, for example, the speaker who stands for Leibniz, Theophilus, suggests a notion of innate truths, which he compares to veins in a block of marble outlining the figure of Hercules which will later be carved out of it, as opposed to a uniform surface or *tabula rasa*. Similarly, 'ideas and truths are innate in us, as natural dispositions, habits or virtualities [*virtualités*] and not as activities [*actions*], although these virtualities are always accompanied by some activities, often imperceptible, which correspond to them'.[30]

The virtual is then a kind of innate tendency towards-force composed of unconscious or low-conscious-level psychological content, on which Leibniz elaborates in his theory of 'petites perceptions'.[31] Leibniz specifies that these virtualities are not necessarily conscious, yet, he writes, they may be swiftly brought to consciousness like the opening line of a song that calls to mind the entire tune.[32]

The term 'virtual' also appears in Leibniz's *Discourse on Metaphysics* (1686), one of the first presentations in consistent form of his metaphysical system.[33] Opposing Descartes and his followers' view of the essence of substance as extension, Leibniz proposes therein instead that the essence of the substance, as noted earlier, consists in its containing all of its predicates. As such, Leibniz redefines the individual substance as a complete individual concept:

> When a number of predicates is attributed to a single subject while this subject is not attributed to any other, it is called an individual substance … Now it is certain that every true predication has some basis in the nature of things, and when a proposition is not an identity, that is to say, when the predicate is not expressly contained in the subject, it must be included in it virtually. This is what the philosophers call *in-esse*, when they say that the predicate *is in* the subject. So the subject term must be included in the subject in such a way that anyone who understands perfectly the concept will also know the predicate that pertains to it.[34]

A substance is defined as that which includes all its predicates, either expressly or virtually. When a predicate is included virtually in the subject, it belongs to its essence; that is, it is included in it *in esse*. Leibniz next moves seamlessly from subject to concept, describing the 'knowable' subject, or the aspect of the subject that may be understood as 'concept', and arguing that perfect understanding of a concept is equivalent to knowledge of all of its predicates. Every concept includes 'traces' of its past and 'marks' of its future, which are fully revealed to

God alone.[35] In order to avoid complete determinism, Leibniz distinguishes, as noted earlier, between 'necessary' and 'contingent' truths, both contained in the concept; the former are 'eternal truths', the contrary of which is a contradiction, while the latter's contrary does not imply contradiction. Contingent truths, which are contained virtually in the concept of the substance, are not necessary but rather based on God's choice of the best amongst several possible options. Leibniz gives the ratios of irrational numbers, which may be solved only through an infinite series, as an example for contingent truths. Thus he can claim that Adam's sin was based on his free action, and not predetermined by God, by arguing that this event was contingent and not necessary.[36]

In his short essay 'On Freedom' Leibniz describes how, at one stage, he had believed that everything in the world was necessary, yet he had changed his mind because of fictitious novels, the possibility of which, as he writes, one cannot deny, although they do not exist in actuality.[37] Deleuze quotes Leibniz's description, in this essay, of virtual inclusion as such that must be extracted from the subject, and as included in the subject 'only with a certain force'.[38] That is, the virtual is included in the subject in a subdued manner, with reduced force, which implies that force is necessary for its actualization.

Against those who argued that ideas are temporary, existing only while we think of something, Leibniz argued for their innate nature. He drew inspiration, as discussed in Chapter 2, from Plato's theory of ideas, referring to his doctrine of reminiscence described in *Meno*.[39] 'This demonstrates that our soul knows all these things virtually and requires only *attention* to recognise truths, and that consequently, it has, at the very least, the ideas upon which these truths depend', writes Leibniz.[40] Knowledge exists in the soul; its actualization requires merely the dedication of attention. The virtual is not opposed to the actual or existing, nor is it equivalent to potential in the Aristotelian sense. The conception of the virtual pertains, as we have seen, both to historical events and to knowledge, a link that is retained in Benjamin's conception of historical knowledge.

2 Virtual histories and infinite totalities

i Natural and virtual history

In Chapter 1 we saw how Benjamin modelled his theory of ideas upon Leibniz's concept of expression, and the way in which the historical aspect of this concept was especially important for Benjamin, leading him to define the essence of the work of art as historical. This becomes even more apparent through Benjamin's usage of Leibniz's term 'virtual' in order to describe a type of history that he distinguishes, in the Foreword to *Origin of the German Trauerspiel*, from what he calls 'pure history'.[41] This distinction is significant not only for the philosophy of history presented in the Foreword but also for Benjamin's later theory of historical experience in the *Arcades Project* and *On the Philosophy of History*. As he writes,

> The fore- and after history [*Vor-sowie Nachgeschichte*] of such essences is, as a mark of their redemption [*Rettung*] or their gathering into the preserve of the world of ideas, not pure but natural history. The life of works and forms – which only in this refuge unfolds clearly and untroubled by human life – is a natural life. If this redeemed being is ascertained to be in the idea, then the presence of the natural-historical fore- and after- history – as opposed to history proper – is virtual. It is no longer pragmatically real but, as natural history [*Naturgeschichte*], is to be read off the state of completion and rest, the state of essentiality.[42]

'Virtual' and 'natural' history are two aspects of the same form of history that is opposed to 'pure history'. 'Natural history' forms part of the essence of the work of art. It consists of the past and future history of the work in virtual, un-actualized form. As in the case of Leibniz's complete individual concept, that includes its history, or the monad, in which past and present are detailed – the history of the work of art forms part of its essential make-up. The 'life of works and forms' can only 'unfold' when not disturbed by 'human life', Benjamin writes, suggesting that works have a kind of 'life of their own', which may only come about if not hindered by subjective assumptions.[43] Virtual history is the same, only applying to ideas rather than to the works of art themselves.

Benjamin establishes links between 'natural history' and 'natural life', a concept he developed in his essay 'The Task of the Translator', to which he refers the reader.[44] Therein Benjamin describes translation as the 'afterlife' [*Nachleben*] of the work, which is in a sense *virtually* contained within the original work, putting forward an objective idea of life virtually detailed in the work.[45]

'The range of life must be determined by the standpoint of history rather than that of nature … [T]he philosopher's task consists of comprehending all of natural life through the more encompassing life of history', he writes.[46] The philosopher is tasked with comprehending the 'life' of works, which can be done only from a historical standpoint. In place of Leibniz's conception of 'living force', Benjamin suggests a concept of forceful life, a 'life' belonging to works of art that unfolds or actualizes under the gaze of the philosopher, and through philosophical analysis. Benjamin appropriates Leibniz's conception of the virtual in order to suggest a notion of history – in this case, the history of works of art – that is essential and objective yet not fully predetermined. It can be 'read' but that does not mean that it has been entirely 'written' in advance.

The terms 'natural history' and 'natural life' also hint to Benjamin's understanding of nature and of the relationship between humanity and nature, and the role played by history in this relationship. While in 'The Task of the Translator' the standpoints of nature and history are opposed, the concept of 'natural history', as defined in the Foreword brings nature and history together by describing history as natural and nature as historical.[47] Doing so, he critiques the early modern dualism between nature and culture and the resulting conviction that nature must be controlled, as he observes in the last section of *One-Way Street*, 'To the Planetarium' [*Zum Planetarium*].[48] There, he points to the early modern period as a turning point in the relation of human society to nature. He suggests that scientists like Descartes, who viewed the mastery of nature for human benefit as the goal of science, transformed the relationship between humans and the non-human environment. Benjamin anticipates in this text ideas on the domination of nature presented by Adorno and Horkheimer in *Dialectics of Enlightenment.*[49]

'Pure history', as opposed to 'natural history', is described as 'pragmatically real' and consisting of an accumulation of past occurrences. Benjamin is probably hinting at progress-driven, historicist accounts of history. On the other hand, virtual history belongs to ideas. 'Redemption' or becoming 'gathered into the world of ideas' is described, then, as bringing about the virtualization of history, rather than its actualization. Benjamin's concept of 'redemption' denotes in this context what he calls 'Platonic salvation', which refers to Plato's theory of recollection discussed in the previous chapter.

Benjamin describes the redemption of phenomena through the recollection of a 'primordial form of perception', and as the recovery of the symbolic aspect of the word through naming.[50] The virtualization of history is brought about by the recovery of 'primordial' knowledge via a process of recollection, or naming, enabling its 'reading' by the philosopher. 'Redemption' is distinct from, and yet is related to, Benjamin's concept of *Erlösung*, used in 'On the Concept of History' and elsewhere, which amalgamates Jewish conceptions of messianic redemption with Christian notions such as apokatastasis.[51]

The 'virtual history' in this case is opposed not to the 'actual' but to the 'real' (unlike Deleuze's concept of virtual, in which the virtual is real). Benjamin writes that the virtual history may be read from the state of completion or rest of the 'essence'; its past and future history may be read, or understood, once it has reached a state of completion, making manifest that the 'virtual' is not oppositional to this state of completion; on the contrary, it is brought about by it. By describing the virtual as becoming 'readable' once the essence reaches a state of rest, Benjamin diverges from Leibniz's account of the virtual force within the substance, which, as we have seen, never reaches such a state of completion or rest, but rather always remains active to a degree. This 'state of rest' will, in Benjamin's later writings, be described as a 'standstill' necessary for the emergence of the dialectical image.

ii Intensive interpretation

Benjamin defines truth as a harmonious relationship between a finite multiplicity of ideas as monads, or self-contained essences, which he likens to stars, or suns: 'just as the harmony of the spheres depends on the orbiting of stars that never touch, so the existence of the *mundus intelligibilis* rests on the ineluctable distance between pure essentialities ... Truth is the sonorous comportment of such essentialities.'[52] Truth is viewed as dependent upon the harmonious distance between ideas, that is, on a musical ratio and balance. Benjamin's reference to the theory of harmony of the spheres, or *musica universalis*, is not accidental. The idea that mathematical and astronomical relations are expressed musically originated in Antiquity and was developed in Medieval philosophy, and seventeenth-century astronomer Johannes Kepler sought to prove it scientifically in his 1619 *The Harmony of the World* [*Harmonice Mundi*].[53]

Benjamin's understanding of truth as non-intentional and in a sense predestined grounds his tasking of philosophers with the 'objective interpretation' of the world, which is disclosed when penetrating deeply into 'everything real' [*alles Wirkliche*].[54] 'In light of such a task of immersion, it does not appear strange that the philosopher of monadology was also the founder of infinitesimal calculus', he writes.[55] He describes Leibniz as a philosopher of immersion and immanence, whose monadology and infinitesimal calculus both follow the same vertical, unidirectional, logic in which the infinite is represented within, and limited by, the finite. Benjamin's references to Leibniz's *Monadology* allow him to criticize the intentional and subjectively structured epistemologies of Kant and Husserl while retaining the individual, fleeting experience as the core of his alternative theory of historical knowledge.

In a letter to his friend and correspondent, the Protestant theologian Florens Christian Rang, Benjamin differentiates between the monadic manner in which the idea is historical, on the one hand, and the manner in which art historians designate a work of art as historical, on the other.[56] The latter type of 'historical' is external, and synonymous with a set of occurrences or events, while the historical nature of the idea is inward-inclined and essential. Benjamin defines these two types of history as extensive 'art history' and 'intensive interpretation'.[57] Intensive interpretation consists of the experience of the viewers, their interpretations or translations of the work, which for Benjamin form part of the work of art itself. Extensive history, on the other hand, does not relate to the works of art, but only takes it as an example. Corresponding to what Benjamin refers to as 'virtual' or 'natural history' in the Foreword, intensive interpretation marks the essential links between works of art.

> For the process of interpretation brings to light connections between works of art that are timeless, yet not without historical dimension. The same forces [*Gewalten*] that become explosively and extensively temporal in the revealed world (that is, history), emerge intensively in the taciturn world (that is, the world of nature and art).[58]

Interpretation illuminates affinities between works that relate to their internal, essential history, rather than the history of the relations between them. These connections are 'timeless' in the sense that they are not successively 'placed' in time.

The distinction drawn by Benjamin echoes what some critics describe as Leibniz's differentiation between inner-monadic and inter-monadic time.[59] The first of these corresponds to his concept of force, or the self-movement of the monad, while the second relates to Leibniz's theory of relational time and space, and his designation of time as a relation of succession between phenomena.[60] Although Benjamin uses the term *Gewalten* rather than Leibniz's *Kraft/la force*, he may have had in mind the self-moving force of monads, the consequences of which become present in the phenomenal world.

Benjamin transforms Leibniz's distinction between the metaphysical world of monads and the physical phenomenal world into a distinction between the world of revelation [*Offenbarung*] and that which is 'taciturn' or sealed [*verschlossen*].[61] Works of art and the ideas that correspond to them are, then, driven by their own self-produced forces, which determine their inner history, becoming explosive when emerging into external history. Benjamin's use of *Gewalten* suggests that the forces internal to the idea, or, as elucidated in his letter to Rang, the work of art, are violent and destructive, even before their emergence into the 'revealed' world. In this respect his conception of force differs from Leibniz's notion of force that contains an entelechy or striving towards perfection.[62]

In the same letter, Benjamin characterizes ideas by their 'intensive infinity' [*intensive Unendlichkeit*]. Ideas are independent essences linked to one another intensively; they are not placed in time in a successive causal relation to one another, and therefore in a sense they are 'timeless', yet each idea has its own history that characterizes its internal development. The infinity of degrees or stages of distinctness in which the idea represents the totality of knowledge, comprising the idea's *Vor-* and *Nachgeschichte*, is included within it as an intensive quality; it is essential and not subordinate to external temporal relations.

iii Violence and emergency: Benjamin, Schmitt, Leibniz

The most significant mark of the mourning play, Benjamin suggests in *Origin of the German Trauerspiel*, is its engagement with history both thematically and formally. As opposed to classical plays, which were based on myths, the subject matter of these German 'mourning plays' consisted of historical events, often depicting political schemes and intrigues, and taking place in royal courts.

The Baroque plays were characterized, as mentioned, by a lack of transcendence, since salvation was no longer possible in the next world, as was the case in classical plays. The situations depicted in the *Trauerspiel*, therefore, had to reach their catastrophic extremes within the play's limits. Benjamin viewed the resulting bloodiness of these plays – comparable to modern-day horror films and castigated by the critical hegemony of his day – as part of their unique reflection of the Baroque worldview. The plots of the plays are described as actualizing, or developing in an infinity of degrees, towards their catastrophic ends. While classical plays offered transcendence, Baroque mourning plays could only counter with the 'liveliest actuality' [*lebenshaftesten Aktualität*].[63]

'If history is secularised in the setting, this is an expression of the same metaphysical tendency which led, in the exact sciences, to the infinitesimal method. In both cases chronological movement is grasped and analysed in spatial images', writes Benjamin.[64] The seventeenth-century Scientific Revolution, and specifically the shift from synthetic to analytic geometry and the introduction of the infinitesimal method, made its mark upon the *Trauerspiel* form. As both Benjamin and Deleuze observe, the introduction of the infinitesimal method into Leibniz's metaphysics resulted in a kind of philosophical immanence, which corresponded with the theological immanence of the Counter-Reformation era. Although Leibniz's system includes an eternal God, it is a timeless God. In other words, he separated infinitely divisible time from a timeless eternity.[65] In the above quote from Benjamin, he argues for a similar immanence in *Trauerspiels*;

since their plots were not structured teleologically, leading towards a redemptive eternity, infinite time and space were condensed into their finite setting. The chronological movement of history was thus abbreviated into images of nature in the stage set design.[66]

The continuous force that dynamizes Leibniz's monad, an immanent force with no outlet, becomes, in Benjamin's reading, a force of violence [*Gewalt*]. Schmitt's theory of the state of exception, an 'extreme doctrine of princely power', as Benjamin describes it, therefore fits well into his reading of the concept of sovereignty in Baroque plays.[67] The ruler holds, according to this doctrine, unlimited power in the case of catastrophe leading to a state of emergency or exception [*Ausnahmezustand*].[68] In Benjamin's reading of Schmitt, he highlights the state of exception as a break, a necessary limitation of dictatorial force rather than its source of power.

'Sovereign is he who decides on the exception' [*Souverän ist, wer über den Ausnahmezustand entscheidet*], Schmitt famously opens his *Political Theology*. The sentence implies a double meaning: the sovereign decides how to react in a state of emergency, as well as what constitutes a state of emergency, in effect creating this state.[69] In his so-called 'decisionism', Schmitt defends the unlimited power of the sovereign, which manifests itself in his ability to declare a state of emergency. Sovereignty is hence defined by exception, with the implication that the foundation of law is in its suspension.[70]

Schmitt and Benjamin both trace the modern theory of the state to political and theological changes that took place during the Baroque. Schmitt writes that 'the central concepts of modern state theory are all secularised theological concepts'.[71] In his eyes the modern state was born from a process of secularization starting in early modernity.[72] This means that for Schmitt, the power of the sovereign replaces that of God in the modern state, and also that he embodies the will of his subjects, to some extent, as the sovereign governs the body politic in Thomas Hobbes' *Leviathan*, or as monads are expressed in God according to Leibniz.[73]

Benjamin recognizes Schmitt's theory of sovereignty as essentially Baroque in its secularized eschatology and its striving for order in a chaotic world.[74] Responding to Schmitt, Benjamin writes that 'the sovereign represents history. He holds historical happenings in his hands like a sceptre'.[75] This sentence bears a double meaning as well. First, within the German Baroque mourning plays, which represent courtly intrigues as stand-ins for historical events, the character of the sovereign represents history. Second, by likening the course of history to a sceptre, symbol of the sovereign's authority, Benjamin suggests that the sovereign derives his status and power from history, but also that his decisions determine its course.

While Benjamin, like Schmitt, finds the origins of the 'state of exception' in the Baroque, he emphasizes these states as the outcomes of catastrophes that result from excessive power of the sovereign. As Benjamin writes, 'in an antithesis to historical idea of restoration, the Baroque is faced with the idea of catastrophe. And it is on the basis of this antithetic that the theory of the state of exception is conceived'.[76] Baroque was an age of conflict and catastrophe, Benjamin writes, as opposed to the Renaissance, a time of stabilization and restoration. In contrasting political catastrophe and states of exception with restoration, which, as we will discuss later on, relates to concepts of eternal return, Benjamin underlines the tensions, within history, between exceptionality or singularity on the one hand and repetition on the other.

In describing Baroque states of exception, brought about by 'war, revolt or other catastrophes', Benjamin emphasizes the circularity of the forces that justifies declaring such states exceptional but, by doing so, extends the power of the sovereign to mark history with more violence.[77] As opposed to Schmitt, who justifies the use of violent means to the end of stable sovereignty, Benjamin critically exposes and critiques the violent origins of law in this section on Baroque sovereignty as well as the earlier 'Critique of Violence' and the later 'On the Concept of History', as Agamben highlights in his reading.[78]

Pondering why Benjamin wrote, in his dedication to Schmitt, that he has derived from Schmitt's later work a 'confirmation of my modes of research in the philosophy of art from yours in the philosophy of state', Weber suggests that they shared a 'methodological extremism'.[79] He quotes the Foreword to *Origin*, where Benjamin describes the idea as 'the formation of the nexus in which the uniquely occurring extreme [*Einmalig-Extreme*] stands with its like'.[80] Weber compares this notion of the unique case that stands at the edge, or extremity of what is familiar, just before it becomes something else, with what Schmitt refers to as the 'borderline case'.[81] Schmitt famously quotes from Kierkegaard's *Repetition* 'the exception is more interesting than the rule', while Benjamin was critical of Kierkegaard's grounding of his ethics in theology.[82] Benjamin and Schmitt may have shared the methodological and aesthetic preference of the exceptional over the usual or the average, yet this should be distinguished from their distinct positions on the political exception or emergency. As we shall see, Benjamin underscores repetition as well as exception through the concept of apokatastasis.

Benjamin transforms Leibniz's immanent concept of infinite force, via Schmitt, into a dialectic between unlimited political force of the sovereign, coupled with indecision and doubt that characterized the depiction of politics in Baroque plays.[83] By reversing the meaning of Schmitt's 'state of exception', Benjamin brings to the fore the necessary limits of power rather than its infinite continuity and exposes the violence resulting from Baroque immanence and underlying Leibniz's principles of continuity and harmony.[84]

iv Image and time-differential

Leibniz's concept of force resonates within Benjamin's theorization of Baroque sovereignty but also in his concepts of allegory and dialectical image, both of which relate to history and historical time. In an early note to the *Arcades Project*, Benjamin describes a 'time-differential' in which the image is real:

> On the dialectical image. In it lies time. Already with Hegel, time enters into dialectic. But the Hegelian dialectic knows time solely as the properly historical, if not psychological, time of thinking [*Denkzeit*]. The time-differential [*Zeitdifferential*] in which alone the dialectical image is real [*wirklich*] is still unknown to him … Real time enters the dialectical image not in natural magnitude – let alone psychologically – but in the smallest gestalt. The temporal momentum [*Zeitmoment*] in the dialectical image can be determined only through confrontation with another concept. This concept is the now of recognisability [*Jetzt der Erkennbarkeit*].[85]

Leibniz defined the differential as the rate of change in the varying quality of matter and motion within the infinite series of degrees of a constant force. The rate, or degree of change of the variable, is derived from the law of its series, determining in turn the inflections of the curves that represent the series. The dialectical image is real, Benjamin writes, only within a framework in which time consists of minute, persistent changes rather than a uniform, stable whole. Hegel's progressive dialectics of negation are contrasted with Leibniz's development as differentiation, a contrast that is also implied in Benjamin's definition of the dialectical process as monadic.[86] Benjamin carefully distinguishes his own articulation of concepts such as 'actualization' and 'dialectic' from that of Hegel. Since, as Benjamin writes, time may only enter the dialectical image through its condensation into the tiniest possible forms, images are particles of history, the totality of which does not consist of an absolute or world history.

As Deleuze writes in *The Fold*, as a preformist, Leibniz understands unfolding, or the development of organic forms brought about by their inner force, as a process in which every fold originates from a prior fold.[87] The debate between epigenesists and preformists during Leibniz's time revolved around the extent to which something is considered as being formed or organized from the 'beginning' or whether organization and form arise over time. From the preformist perspective of Leibniz, genesis is viewed not as a fixed beginning but as the process of becoming over time. The unfolding of a series, just like the development of an organic form, is, according to Leibniz, a kind of organization of existing form, or actualization through differentiation.

In the 'now of recognisability' in which two times are brought together, such actualization through differentiation is performed. The image does not form part of dialectical time; rather, time enters the image and is compressed into its 'smallest gestalt'.[88] The image is dialectical in that it includes an unresolvable tension between enfolding and unfolding, actualization and its inverse process, a tension motivated by momentum or force. Time is condensed into the spatial form of the image only to be actualized once the image is read and recognized.[89]

An early sketch for the *Arcades Project* is illuminating in respect to the relation between temporality and image:

> Historical knowledge of the truth [*Geschichtliche Wahrheitserkenntnis*] is possible only as the overcoming of semblance [*Aufhebung des Scheins*]. Yet this overcoming should not signify sublimation, actualisation of the object but rather assume, for its part, the configuration of a *rapid* image. The small quick image in contrast to scientific complacency. This configuration of the rapid image goes together with the acknowledgment of the 'now' in things. But not the future.[90]

The 'rapid image' is defined as the alternative to sublation in a manner that foreshadows Benjamin's later description of the dialectal image as containing conflicting tensions. The overcoming of semblance in order to reach knowledge of the truth is performed through a quick, small, sketch-like image of the historical object rather than traditional scientific methods. The description of the image as 'small' is reminiscent of the monad's abbreviation or reduction of the world into its minute dimensions.

3 Mallarmé and Benjamin on chance

i Attested auditor

The term 'now of the knowability', which, as discussed in Chapter 1, first appeared in a fragment dated to 1920–1, carries with it a danger of historical predestination that Benjamin sought to avoid. This struggle is also evident in the fragment 'Attested Auditor of Books' [*Vereidigter Bücherrevisor*], published in *One-Way Street* (1928).[91] In this text, which focuses on Mallarmé's poem *Un coup de dés jamais n'abolira le hasard*, Benjamin draws a connection between the 'intensive' nature of the image and foresight into history. Mallarmé is praised for having read the true image of the future: 'Mallarmé, who in the crystalline structure of his manifest traditionalist writing saw the true image of what was to come, was in the *Coup de dés* the first to incorporate the graphic tensions of the advertisement in the printed page.'[92]

The struggle between contingency and chance, on the one hand, and human action, on the other, lies at the heart of the poem's thematic. Specifically, the poem engages with the manner in which chance and necessity affect literary creation, as is evident in lines like 'Choit/ la plume/ rythmique suspense du sinistre' ['Down falls/ the plume/ rhythmic suspension of disaster'].[93] The struggle remains unresolved, the oblique narrative of the poem, like the maître himself, hovering in the hypothetical and the recursive, as we can see for example in the graphic repetition of the words 'come si' and 'si', and the likening of the master to a shipwrecked prince – a reference to the famously indecisive Hamlet. The struggle between contingency and chance is thus left unresolved, as accentuated by the many repetitions of words, such as '*Un coup de dés*', words that both open and close the poem creating a circular structure.[94]

Benjamin suggests that in Mallarmé's poem, the relation between truth, on the one hand, and image, on the other, is articulated through the *Bildlichkeit* (imageability or image-potential) of the text, which may only be achieved in poetic form. By writing a poem in which the text itself creates images, through its formatting on the page, Mallarmé had pre-empted the modern reconfiguration of script-image relations.

If in fact, as Scholem recalls, Benjamin felt that he did not understand *Un coup de dés*, this did not impede him from highlighting in his text a key aspect of the poem, as Mallarmé viewed it – namely its foreshortening technique that condenses an infinite undecidability into succinct formulations and imagery.[95] By creating a unique *Schriftbild*, or image-writing, Benjamin suggests, Mallarmé anticipated the new technical and literary forms that will come to replace the book (ironically, since Mallarmé's unfinished work, simply titled 'The Book [*le Livre*]', was meant to be an all-encompassing 'book of books').[96]

'Everything takes place in foreshortened, hypothetical form', Mallarmé writes in the preface to the poem.[97] These lines might have struck a chord in Benjamin, since, as discussed, he describes the idea as a monad that includes an image of the world in condensed or foreshortened form in the *Trauerspiel* book, and this expressive structure reverberates in his later concept of the dialectical image.

ii International moving script

> *The typographic experiments later undertaken by the Dadaists … show the contemporary relevance of what Mallarmé, monadically, in his hermetic room, had discovered through a preestablished harmony with all the decisive events of our times in economics, technology, and public life.*[98]

The image of pre-established harmony that Benjamin uses in 'Attested Auditor' is borrowed from Paul Valéry (1920), echoing Leibniz.[99] When describing pre-established harmony, as mentioned, Leibniz invokes the metaphor of musicians playing in separate rooms, yet reading from the same sheet of notes, which ensures that the multiple instruments produce harmonious music.[100] Valéry may have derived the image of harmony from Mallarmé himself, who compared the agony of poetic creation with pure harmony in his lecture *La musique et les lettres*.[101]

On the one hand, broken off from the external world, Mallarmé followed his internal stylistic inclinations, and, on the other, he formed a continuity with the world outside, through the reflection of the world's constantly transforming totality within the poem's images. The poem is *bildlich*, then, in the manner in which it consists of an image of its present and contains its future in virtual form. The *Schriftbild* created by Mallarmé, suggests Benjamin, condenses future events into words scattered in isolation upon the double pages of *Un coup de dés*.[102] By outlining a relation of expression between the poem and the exterior world, Benjamin highlights the foresight of poetic creation in its distinctive use of images, through which it provides singular representations that condense a totalizing image of world history. As described by Meillassoux, in Mallarmé's poem there is 'a dialectical infinite … that includes its other … a non-Hegelian dialectic, one without progress'.[103]

Benjamin suggests, furthermore, that through its radical use of written-image [*Schriftbild*] to reflect its historical moment, Mallarmé's poem becomes politically effective. It precipitates the next revolution, writes Benjamin, by internalizing the colourful, chaotic lettering of advertisements.[104] The liberation of script is described as a process in which it becomes increasingly closer to the image: the shift from the vertical plane of the book to the perpendicular of films and billboards leads up to the three-dimensionality of the card index.

Benjamin defines the card index [*Kartothek*] as a modern-day rune, in which, like a monad that mirrors the world, an isolated word stands for the full text.[105] His fascination with the index-form continued in later years as manifest, for example, in the thematic lists composed as part of the drafting of the Baudelaire essays, in which entire passages are abbreviated into themes such as 'allegory' or 'boredom', each marked by a colourful symbol or image.[106]

In the concluding paragraph of 'Attested Auditor', movement, or time, is implied as a future fourth dimension of script:

> In this picture writing, poets, who will now as in earliest times be the first and foremost experts in writing, will be able to participate by mastering the fields in which (quite unobtrusively) it is being constructed: statistical and technical diagrams. With the founding of an international moving script, poets will renew their authority in the life of peoples.[107]

Poets, masters of exercising the eccentric imageability [*exzentrischen Bildlichkeit*] of script, will lead the way in the construction of a universal, ever-fluctuating script that will capture in diagrams and statistics the 'life of peoples': their behaviour and moral judgements. Benjamin alludes to the *characteristica universalis*; the universal language Leibniz imagined would reduce all natural language to pictographic characters that would depict reality as mathematical notations reflecting numbers and magnitudes.[108] Judgements of reason, Leibniz conjectured, could be easily made in such a language via the calculations of probabilities, eliminating the chance occurrences that Mallarmé's throwing of the dice could not overcome.[109] As noted, Benjamin rejected the attempt to reduce all that is expressed in language into logical notation. His reading of Mallarmé's poem goes against efforts at deciphering its 'code'. Benjamin seemed to believe, like Deleuze, that the mysterious number at the heart of *Un coup des dés* was contingent.[110]

Benjamin's 'international moving script' rather offers a vision for a merging of script and image that would go beyond Mallarmé's *bildlich* poem in its abbreviation of human history, perhaps taking the form of musical notes or cinematic moving-images. Hovering in the hypothetical, the poetic expression of Mallarmé leads the way. The words of his poem, separated by the blank spaces between them, each condense a totality of human experience reflected from a different point of view, strewn upon the pages like stars in a constellation. In Mallarmé's words, 'Rien/ n'aura eu lieu/ que le lieu/ exepté peut-etre/ une constellation' ['Nothing/ will have taken place/ but the place/ except perhaps/ a constellation'].[111]

iii Thought-worlds

In the third chapter of *The Fold*, 'What Is the Baroque', Deleuze suggests that the fold was Mallarmé's most important operational act, which made him a 'great baroque poet'.[112] Mallarmé's poems enact a process of 'folding' in which exterior and interior are distinct yet connected, holding a tension in which each fold contains the next.[113] They thereby embody, for Deleuze, the key Baroque concept of the fold that 'unfurls all the way to infinity'.[114]

However, at a later stage Deleuze returns to his own comparison of Mallarmé and Nietzsche from *Nietzsche and Philosophy*, painting a more complicated picture. Mallarmé and Nietzsche are described here as being followers of Leibniz in their creation of 'thought-worlds', and yet at the same time, and in opposition to Leibniz, they are fellow nihilists. 'For them the world lacks principle, has lost its principles. That is why the roll of the dice is the power of affirming chance.'[115]

Mallarmé and Nietzsche choose a world in which principles are absent, as opposed to Leibniz's proliferation of principles, designed to bridge the gaps created during the lengthy political and theological crises of the Baroque.[116] As such, Mallarmé and Nietzsche were visionaries of modernity, creators of a multiplicity of 'thought-worlds' rather than the multiple principles of the Baroque. Chance is here opposed by Deleuze to the logic of principles and the overdetermination of meaning of Leibniz's world, in which everything has a reason.

In *Nietzsche and Philosophy*, Deleuze emphasizes rather the differences between Mallarmé and Nietzsche's 'throwing of the dice'. For Mallarmé, Deleuze writes, chance and necessity were on opposite sides, each of which denying the existence of the other.[117] The success of the dice-roll is measured by its ability to eliminate chance. For Nietzsche, by contrast, throwing the dice affirms both chance and necessity: when thrown in the air, the dice affirms chance, while when landing and revealing a certain number combination, necessity is affirmed.[118] This understanding of the inseparability of chance and necessity guides Nietzsche's conception of 'eternal return' – 'the necessary result of all chance' – in Deleuze's reading.[119]

4 Spectres of eternal returns

i Apokatastasis

A Greek term designating restoration, and used to describe the Stoic cycle of cosmology, 'apokatastasis' was adopted by Christian theologians, most notably Origen of Alexandria, who used it to describe the restoration and salvation of all souls. The term therefore has both a Stoic and Christian lineage, the latter based on an apostolic reference to Jewish eschatology, namely St Peter's speech in Acts 3.21, describing Christ 'whom heaven must hold until the times of that restoration of all things (achri chronōn apokatastaseōs pantōn)'. It follows that apokatastasis is interlinked with messianic traditions, both Jewish and Christian. The return of all things is described as a universal restitution following the return of the Messiah.

Benjamin uses the term apokatastasis in different texts written in his last years. In the *Arcades Project*, he describes apokatastasis as part of a 'methodological proposal for the cultural-historical dialectic'.[120] Therein and in notes for his last text 'On the Concept of History', Benjamin also discusses eternal return in relation to Nietzsche and Blanqui.[121] Yet his concept of apokatastasis, although recently receiving some attention, is not often linked to his reading of Leibniz.[122]

Hegel's dialectic, on the one hand, and Leibniz's infinite differentiation or variation, on the other, are once again pitted against each other in Benjamin's 'Convolute N':

> Modest methodological proposal for the cultural-historical dialectic. It is very easy to establish oppositions, according to determinate points of view, within the various fields of any epoch (…) The very contours of the positive element will appear distinctly in so far as this element is set off against the negative. On the other hand, every negation has its value solely as background for the delineation of the lively, the positive. It is therefore of decisive importance that a new partition be applied to this initially excluded negative component so that, by shifting the angle of vision [*Verschiebung des Geschichtwinkels*] (but not its scale [maßstäbe]!) a positive element emerges in it too – something different from that previously signified. And so on, ad infinitum, until the entire past is brought into the present in a historical apocatastasis.[123]

Benjamin suggests a 'cultural-historical dialectic' in which a process of infinite differentiation will take place, instead of negation and *Aufhebung*. Breaking the negative into components and shifting one's point of view opens up a view of the positive within the negative, he suggests, and this repeated procedure results in a 'historical apocatastasis'. The changing of one's angle of vision and its attendant revelation of hidden truths is a recurrent idea in Benjamin's writing, for example in the opening of 'The Storyteller'.[124] Leibniz's perspectivism – according to which the totality of truth, accessible only to God, is composed of the infinity of points of view of the infinity of monads – underlies Benjamin's recurrent designation of a multiplicity of points of view as constituent of truth.

When mentioning apokatastasis in 'The Storyteller', Benjamin refers to Origen's doctrine of universal salvation from *De principiis*, describing it as the 'entry of all souls into paradise'.[125] Benjamin's conception of redemption [*Erlösung*], famously utilized in 'On the Concept of History', is understood by Löwy as referring to both the concept of apokatastasis and Jewish messianism.[126] Both religious concepts, he writes, share a double meaning: in apokatastasis, the restitution of the past is at the same time a renewal.[127] The same dual meaning, restoration of the past on the one hand and a utopian inclination on the other, resides in the Jewish Messianic tradition, as described by Scholem.[128]

Benjamin expressed a similar idea in a tale he recounted in a conversation with Ernst Bloch, recorded by Bloch and quoted by Giorgio Agamben, who neglects to mention that the source of the so-called Hassidic tale may have been no other than Scholem:

> The Hassidim tell a story about the world to come that says everything there will be just as it is here. Just as our room is now, so it will be in the world to come; where our baby sleeps now, there too it will sleep in the other world. And the clothes we wear in this world, those too we will wear there. Everything will be as it is now, just a little different.[129]

What is specific about Benjamin's version of the story, writes Agamben, is the 'tiny displacement that the story introduces into the messianic world'.[130] Benjamin's version of historical apokatastasis or redemption describes a possibility of infinite repetition and differentiation rather than a utopic final completion. There is no reason to assume that Benjamin had read Leibniz's unpublished drafts on 'apokatastasis', where he describes it as the eternal return of history.[131] However, Leibniz expresses similar ideas in brief in the *Monadology*, to which Benjamin refers, and, moreover, Origen is a shared source for Benjamin and Leibniz's use of the term.[132]

In 'Apokatastasis Panton' from 1715, Leibniz argues that since the number of books on earth is finite, and public annual history on earth can be related in a book of a certain length, if humanity were to last long enough in its current state, public histories would necessarily be exhausted and, therefore, they would start repeating themselves. Leibniz suggests the narrated accounts of events will return, or in other words the repeated events will warrant the same accounts, although there may be slight differences between the repeated sets of events themselves.[133] The slight variations within repeated accounts allow for the progression of history.[134] Leibniz refers, in his essay, to 'the great

Platonic year' – the idea presented by Plato of a 'complete' or 'perfect' year, achieved when all eight celestial revolutions or periods return to their starting points.[135]

Benjamin uses the conception of apokatastasis to describe imperfect repetition, yet forgoes the progressive or perfection-inclined aspect of the traditional use of the term, as demonstrated by the Hasidic story in which the world to come is described as a slightly unhinged version of the earthly one. Actualization as differentiation to infinity, transformed by Benjamin into what may be defined as non-Hegelian dialectics, is at work in the conception of apokatastasis as within the dialectical image. In these two structures, a similar condensation of time past into the present is performed.

The infinite differentiation within the image goes against the traditional identification of image with likeness, drawn from Gen. 1.26, where it is written that man is created 'in the image and likeness' of God. It also contests Plato's notion of the *eikon* as 'faithful resemblance'.[136] In Benjamin's 'image', the relationship between genesis and essence, which he had challenged in the *Trauerspiel* book, is questioned once again: essence is not determined through genesis, as in the case of the creation of man. Rather, it is determined by a process of differentiation: not one of likeness, on the one hand, or negation, on the other. Creation results from a repetition of the same while shifting one's angle of vision, or minutely deferring one's moment in time.

ii Modern returns: Blanqui and Nietzsche

Deleuze and Benjamin are both drawn to expression, and to eternal return as its spatial equivalent. Figures of eternal return have dominated philosophy in both antiquity and modernity. As discussed, in his elaboration of the concept of apokatastasis, Leibniz alludes to Plato's theory of return. Deleuze refers, in *Nietzsche and Philosophy*, to Nietzsche's distinction between his own figure of eternal return and notions of cyclical return in antiquity. Benjamin, however, viewed Nietzsche's return as continuing a mythical approach: 'Thinking the idea of eternal recurrence once more in the nineteenth century, Nietzsche becomes the figure on whom mythic doom is now carried out.'[137]

Deleuze follows Nietzsche in describing ancient and modern figures of return as opposed; in the first, he writes, time is subordinated to a cyclical, eternal movement of the world.[138] The eternal cycle is predestined, resulting from divine decision. In the second, modern version, time is no longer subordinated to movement and the cycle results from blind chance or necessity.

Despite noting the continuity between these different forms, Benjamin discusses 'Modern' figures of return put forward by Louis Auguste Blanqui and Nietzsche, separately from such notions as apokatastasis and redemption. Unlike Deleuze, who adopts a form of eternal return, Benjamin is critical of these visions of return, describing them as nightmarish, yet also as intensely powerful.

In the second Exposé he wrote for the *Arcades Project* in 1939, Benjamin delves into Blanqui's vision of eternal return, presented in *Eternity by the Stars* (1872), which he seems to prefer over Nietzsche's *Zarathustra*, to which he refers only briefly.[139] Benjamin describes Blanqui's doppelgänger world of 'present eternalised', in which every component of the universe is repeated to infinity, as the 'phantasmagoria of history itself'.[140]

By using the term 'phantasmagoria' he evokes the nineteenth-century horror theatre, which used magic lanterns to project images of ghosts, skeletons and other haunting images. Phantasmagoria is mentioned explicitly in another *Arcades* fragment as part of a list of nineteenth-century pre-cinematic spectacles produced by emerging visual technology.[141] He casts Blanqui's 'return' as transforming history into an alarming spectacle through the use of visual effects, and comments on its 'extreme hallucinatory power'.[142]

Benjamin also refers to Marx's pronouncement of the phantasmagorical powers of the commodity, which Benjamin links, in the *Arcades*, to a manifold of visual technology, among which 'phantasmagoria' has a privileged place, with tropes of ideological illusion.[143] Benjamin describes the return of history as its commodification: 'the idea of historical recurrence transforms the historical event itself into a mass produced article.'[144] As a process of commodification of history, eternal return brings about the decline of its 'aura', historical events losing their unique status in time.

This phantasmagorical figure of eternal return is also described as a figure of Baudelairean modernity.[145] As such, it is to be distinguished from the Baroque and its allegory, the latter described by Benjamin as 'eternal transience'.[146] Benjamin does not hide his disappointment in Blanqui, writing that 'this resignation without hope is the last word of the great revolutionary'.[147] As Benjamin acknowledges, although Blanqui's vision included a critique of historical progress, he ultimately chose return over progress.[148]

For Benjamin, both progress and eternal repetition of the present are not acceptable models of history. As he writes, 'it is the inherent tendency of dialectical presentation to dissipate the semblance of sameness, and even repetition, in history. Authentic political experience is absolutely free of this semblance.'[149] Benjamin's dialectical model of history as montage consists of his 'third way' in which historical time is neither linear nor circular – but constructed and expressive, as will be discussed further in Chapter 5.

If, for Benjamin, the eternal return of history consists of a 'vision of hell', for Deleuze, Nietzsche's 'eternal return' consists of a 'Copernican Revolution'.[150] By furnishing difference with its own concept, Nietzsche's eternal return was the precursor of Deleuze's own revolutionary philosophy of difference.[151] Deleuze describes Nietzsche's eternal return, in *Difference and Repetition*, as a return not of the 'identical', the 'same' or the 'One'. Rather, 'returning constitutes the only Same of that which becomes. Returning is the becoming identical of becoming itself'.[152] Eternal return offers a model in which becoming takes priority over being. Starkly distinct from both the Ancient Greek return as law of nature, on the one hand, and Blanqui's dark vision, on the other, Nietzsche's eternal return, in Deleuze's reading, does not produce infinite copies of the same. It produces repetitions and selects differences in an ever-turning wheel of becoming.[153]

As noted in Chapter 2, at times Deleuze reads Leibniz through Nietzsche. In Deleuze's discussion of Nietzsche's eternal return, as described in *Difference and Repetition*, we see that the inverse is true as well. Difference is not a term used by Nietzsche, and, as Deleuze concedes, Deleuze's own conception of 'difference in itself' was inspired by Leibniz's differential. What is more, Deleuze highlights the principle of selection, an important one for Leibniz that determines the selection of the best possible world, within Nietzsche's eternal return, in order to explain how the return is a principle of difference. Accordingly, he describes how only the 'extreme and excessive', or in other words the different, within the 'theatrical' world of eternal revolution, returns.[154] While in Leibniz's system it is God who selects the best, actual world from among many, in Deleuze's reading of Nietzsche, eternal return serves in itself as a principle of selection. Deleuze describes eternal return as a reformulation of Kant's categorical imperative, in which universalization is replaced with repetition: 'whatever you will, will it in such a way that you also will its eternal return.'[155]

In the fifth chapter of *Difference and Repetition*, 'Asymmetrical Synthesis of the Sensible', the link between Nietzsche's eternal return and Leibniz's metaphysics and dynamics becomes clearer, when Deleuze writes that 'the eternal return is neither qualitative nor extensive but intensive, purely intensive'.[156] The differences produced by returns are variations in intensity, degrees of power.

Through Nietzsche's eternal return, Deleuze proposes a new way of thinking time, in which difference of intensity replaces hierarchization. This is the ethical underpinning of eternal return, which expresses, he writes, the common being of all things as that which returns.[157] By highlighting the intensive nature of eternal return, Deleuze picks up on Nietzsche's Leibnizianism. As Deleuze writes, Nietzsche's interest in contemporary energetics was not guided by mere nostalgia.[158] Leibniz's conception of force was important for Nietzsche, as is clear, for example, in Nietzsche and Leibniz's respective understandings of all events being governed by immanent internal forces.[159] Both in itself and via Nietzsche's eternal return, Leibniz's force was also crucial for Deleuze's formation of the category of 'intensity', definitive for his understanding of both being and experience. Intensity, he writes, forms part of the conditions of real experience, its variations determining the being 'of' the sensible.[160]

Although Zarathustra is a teacher and prophet-like figure, in Deleuze's interpretation, the becoming [*devenir*] of eternal return is not oriented towards a future [*avenir*] of messianic expectation nor of capitalist progress. Return does not lead to perfection, salvation or destruction. It simply affirms becoming, thereby posing a challenge to dialectical and transcendental thought. Deleuze's model of return is therefore unlike the one Benjamin refers to in many ways, yet they share the critique of progress-driven thought and history, as well as supporting an anti-historicist, unhinged, interrupted temporality.

Although neither refers directly to Leibniz's late reflections on eternal return, Deleuze and Benjamin link Leibniz's spatial structure of expression, which allows the inclusion of multiple, virtual worlds within our own, with the structure of eternal return. Their readings make manifest that what is at stake in Leibniz's reflections on eternal return is the inherent repeatability of events rather than their actual repetition.

While in Deleuze's philosophy, concepts of force and virtuality appear often and explicitly, in Benjamin's these concepts and other 'vitalistic' elements are the subject of much debate. I have shown, however, in this chapter that Leibniz's monad, as an expressive, intensive substance endowed with force is a significant concept in Benjamin's philosophy of history, in earlier and later texts, and relates, as well, to his conception of eternal return. Benjamin's use of the structure of expression and its temporal equivalent in eternal return links his reading of Leibniz to that of Deleuze.

Having mainly explored Leibniz-inspired concepts of force, virtuality and eternal return in this chapter, in the next I will focus on ways in which Baroque perspectivism informed Benjamin and Deleuze, especially in their concepts of allegory and Mannerism. We will see how Leibniz's concepts of expression and perspectivism inform their conceptions of temporality, politics and art.

4

It's all about perspective: The body politics of the Baroque image

Introduction: Phantom images and perspectival devices

This chapter interrogates how the revolutionization of perspective that took place during the Baroque informed Leibniz, and in turn, Benjamin and Deleuze. It focuses on the role of perspective and image within Leibniz's theories of knowledge and perception and the part these concepts, and Leibniz's perspectivism more generally, play in Deleuze and Benjamin's conceptualizations of the image and vision more generally.

The chapter outlines links between Leibniz's research in optics and his philosophical perspectivism. In general, philosophical perspectivism argues for the contingency of truth in relation to the multiple perspectives of individuals. In some versions, such as Whitehead's, it argues that the world itself is reducible to the perspectives of subjects upon it.[1] Its proponents are therefore often accused of relativism, a charge Leibniz attempts to avoid. Perspectivism has enjoyed a recent revival in philosophy of science, yet it is seldom discussed outside of this domain.[2] Both Benjamin and Deleuze, I argue, assimilate versions of Leibniz's perspectivism, albeit in very different ways. Benjamin broaches what might be called a 'historical perspectivism', while Deleuze's interpretation of Leibniz in *The Fold* results in what can be referred to as a 'mannerist perspectivism' that converges with his thought on cinema.

According to Leibniz's perspectivism, the individual is constituted through his or her points of view, through which it is differentiated from all other individuals. Leibniz refers to these multiple 'points of view' as 'perspectives', hence we can describe his conception of individuality as determined through perspective. Leibniz's understanding of expression, and hence of perception, draws on the work of contemporary mathematicians such as Girard Desargues and Blaise Pascal.[3]

As highlighted by Deleuze, the individual point of view is linked to the monad's body: the distinct zone of expression of the monad, the extent to which its perceptions are clear, established by its point of view, is coterminous with the body. As discussed in Chapter 1, perception is the term Leibniz uses when discussing the psychological aspects of the projective geometry-derived structure of expression.[4] The monad's perception of the world depends on its point of view and therefore, in a sense, perspective is a constructive aspect of perception.

These aspects of Leibniz's perspectivism are crucial for Benjamin and Deleuze. Specifically, they draw on the manner in which Leibniz's perspectivism is constitutive of individuality and on his theory of consciousness and 'minute perceptions'. As already observed, both Deleuze and Benjamin pay special attention to Leibniz's emphasis on hallucinatory or intoxicated states of perception, understanding perception as constitutive of the body through differentiation. Benjamin and Deleuze utilize and develop key elements of Leibniz's metaphysics within which vision, perception and body are linked through a theory of 'point of view'.

Leibniz's perspectivism, read through the prism of Mannerism, forms part of Deleuze's ongoing critique of image and artifice, manifest in his attack on 'the image of thought' and his conception of the simulacrum. As we will explore in more detail in Chapter 5, these concerns form part of Deleuze's image theories in the two *Cinema* books, on which he worked while delivering his series of lectures on Leibniz.

1 Leibniz's conceptions of image and perspective

i Image

Perspectival vision has often been used by philosophers as a central metaphor for theories of knowledge and consciousness, and Leibniz was no exception.[5] He was unique, however, in drawing upon contemporary theories in projective geometry and perspective, as well as his own work in infinitesimal calculus, for his theory of perception. His understanding of perspective was also shaped by a long-lasting engagement with optics and visual apparatuses.[6] In an essay from 1696, Leibniz discusses reflection involving concave mirrors, suggesting that his recurrent 'living mirrors' metaphor was inspired by his research on reflection.[7] Benjamin and Deleuze, as I have suggested, were informed by these concerns, as well as by other aspects of Leibniz's metaphysics.

In *New Essays on Human Understanding*, Leibniz uses the term 'image' repeatedly.[8] The work, which responds to Locke's *An Essay Concerning Human Understanding* (1690), is structured as a dialogue between Philalethes, spokesman for Locke, and Theophilus, Leibniz's own spokesman.[9] Leibniz attempts to refute Locke's understanding of the human mind as a 'blank slate'.[10] As discussed, Leibniz himself believed, to the contrary, in the existence of innate ideas. In *New Essays*, the 'image' denotes a representation of a sensible quality that either willingly or unwillingly affects our senses.[11] In several passages, Leibniz describes images as 'impressions':

> But a single very strong impression may, by its very intensity, instantly have as much effect as could be had by a repetition of mild impressions over a long period of time; and so this intensity may etch into the imagination [*fantasie*] as deep and vivid an image as prolonged experience produces.[12]

The image is described as etched or marked into imagination. An intense image can mark the soul immediately, while less intense ones may have a comparable effect over time. In other passages Leibniz describes dream-images and images linked in memory.[13] Like colour, the image can be more, or less, intense; Leibniz describes the image as equivalent to sense experience such as 'colours, taste and so on'.[14]

Images have a degraded status, since they come to us through confused perceptions, and may therefore be illusionary, or merely phantoms ['*Ne sont que de fantômes*'], as Leibniz writes.[15] Furthermore, Leibniz describes how images may bring about error, by producing a false connection between two ideas within the imagination, so that one erroneously judges the connection between them as causal.[16] Speaker Theophilus, who represents Leibniz's own views, gives the following example: when we look at a rotating wheel in motion, our confused image [*fantôme*] of it makes its teeth disappear and an imaginary ring appears in their place.[17] Our visual perception, in other words, should not be trusted to truthfully depict the reality before us. In this regard, it is easy to understand why the terms 'image' and 'imagination' (Leibniz uses both *imagination* and *fantasie*) are often used in proximity. Thus, for example, Leibniz describes the imagination of animals as an activity of passing from one image to another.[18]

In the second book of *New Essays*, Leibniz denounces Locke for conflating the distinction between *idea* and *image*. Theophilus gives the following example: It is usually impossible to determine that a polygon, a plane figure made of line segments, is a chiliagon, which is a polygon with a thousand sides, just by looking at its image, without counting. The image of the chiliagon presented to the senses cannot supply the correct data that would determine its nature as a chiliagon.[19] The 'clear image' one may have of these shapes consists merely of a 'confused idea' of their nature and properties.[20] Although the image of the shape is clearly visible, our ideas of it are confused until supplanted with mathematical information. Thus both image and idea are representational, the difference between them consisting in what aspect of the thing they represent: the image represents the sensible qualities of the thing, while the idea stands for its essential qualities.

Leibniz's use of the term 'image' is consistent with how it was used within the idealist philosophical tradition. Plato's conception of images, specifically the idea that images that are founded upon sense experience lead to illusion and error, is manifest in Leibniz's usage. Leibniz uses both '*image*' and '*fantôme*', the latter an 'erroneous' image, which echoes Plato's distinction between *eikon* (semblance) and *phantasma* (illusion).[21]

The notion of an image as impressed upon the soul or in memory, and preserving the past in the present, also present in Plato, was hugely influential on European philosophy.[22] Leibniz's distinction between image and idea, and his use of the broader conception of imagination, perhaps also drawing on Plato and Neo-Platonism, may have pre-empted the special place assigned to imagination by Kant.[23] The conception of image within the idealist tradition, as represented by Plato and Leibniz, is one that both Deleuze and Benjamin repeatedly and critically challenge in their writings.

ii Perspective

Aside from his notion of the image, I have spoken of Leibniz's own attention to theories of perspective, a recurring theme in a variety of his works. Leibniz repeatedly uses the terms 'perspective' and 'point of view' and took an active part in the study of lenses, as manifest in a letter addressed to him by Spinoza, which responds to Leibniz on the advantages of a specific type of lens.[24] During his time, 'perspective' was used, as it is today, to denote the field relating to the representation of three-dimensional objects on a two-dimensional surface.[25]

The unique aspect of Leibniz's theory of perspective lies in its relation to individuality. Not only is the individual's perception constituted by its point of view but the individual itself is thereby constituted, and therefore individuality may be said to be, in a sense, perspectively determined. According to Leibniz's theory, the substance or individual is constituted by its point of view on the world, which is different from all other points of view of all other individuals.

Each simple substance is thus rendered individually thanks to its singular point of view. Leibniz illustrates this using the recurrent metaphor of alternate perspectives of a city, for example in the *Monadology*:

> Just as the same city viewed from multiple directions appears entirely different and, as it were, multiplied perspectively … because of the infinite multitude of simple substances, there are, as it were, just as many different universes, which are, nevertheless, only perspectives on a single one, corresponding to the different points of view of every monad.[26]

Since an infinity of monads exists, each with a unique point of view, there is an infinity of perspectives that constitute the entirety of the universe, mirroring God's infinity. The conception of a perspectivism, which designates individual monads as constituted by their points of view, allows the single universe to comprehend a multiplicity of different perspectives, or 'universes'.

The word 'perspective' comes from *perspicere*: to look through, into or at, as well as to perceive clearly. During the seventeenth century in France, a 'very creative period in the history of perspective', developments in perspective went hand in hand with those in projective geometry.[27] Leibniz had access to at least some of these developments, for example through Pascal's *Traité sur les coniques* (1654) and Desargues' *Brouillon Project* (1636), whose work he mentions in *New Essays*.[28] The engineer Desargues (1591–1661) developed his manual for the construction of perspective *La Perspective* (1636) and, in the second part of the *Brouillon Project*, focused on conic sections, a beloved theme of Leibniz.[29]

Leibniz mentions Desargues' work in *New Essays* in the chapter 'Of Perception'. He uses perspective as an example for the deceptions of painting, which leads us to confuse image for reality and substitute cause for effect.[30] By applying Desargues' methods to painting, Leibniz writes, the creation of three-dimensional representations using hue and shading was enabled. 'Flat' two-dimensional depictions of objects painted 'in the Chinese manner' were replaced by shapes rendered with shading, which made possible the distinction between a spherical surface and a circular one.[31]

The gradual shift from darkness to light in paintings that use perspective is employed by Leibniz as a model for his theory of knowledge in which there is a gradated transition between distinct and confused ideas. This distinguishes his theory of knowledge from Descartes', in which the distinction between distinct and confused ideas is clear-cut.[32] Beyond referring to Desargues' methods of shading, Leibniz also advanced his own 'doctrine of shading' which he viewed as a kind of reverse perspective, in which a source of light replaces the position of the eye.[33]

In addition to his methods of shading, Desargues' work was important for Leibniz's research into conic sections. Debuiche argues that one of the ways in which Desargue and Pascal's theories of perspective, and specifically their understanding of conic sections, influenced Leibniz was in their use of the term 'correspondence'. Projective perspective, she writes, functions as a transformation that preserves the relations between two elements and thereby constructs a relation of analogy between them.[34] These relations serve as a model for Leibniz's concept of expression.

In his *Theodicy* Leibniz makes a direct connection between conic sections and the relations of analogy they construct, on the one hand, and the monad's representation of the universe from its unique point of view, on the other, which, we might recall, is equivalent to how the soul perceives the universe:

> It is true that the same thing may be represented in different ways; but there must always be an exact relation between the representation and the thing, and consequently between the different representations of one and the same thing. The projections in perspective of the conic sections of the circle show that one and the same circle may be represented by an ellipse, a parabola and a hyperbola, and even by another circle, a straight line and a point. Nothing appears so different nor so dissimilar as these figures; and yet there is an exact relation between each point and every other point. Thus one must allow that each soul represents the universe to itself according to its point of view, and through a relation which is peculiar to it; but a perfect harmony always subsists therein.[35]

Conic sections, obtained by projecting a circle upon a plane that 'slices' the cone, demonstrate how one and the same circle may be expressed by various shapes, each of which seems entirely different and yet retains the analogical relation between each of the points in the expressed circle and those in the expressing shapes.[36] Leibniz likens this analogical expression to how every soul represents the universe, even while an eternal harmony subsists between their representations.

Leibniz uses examples from perspective, then, to illustrate both expression and perception. As Serres writes, the relation, within conic sections, between the multiple curves that may be projected upon one point comprehends the relation of the one and the multiple. They can therefore serve as a model of expression in general and of perception in particular.[37]

Another example may be found in a letter to Des Bosses, in which Leibniz distinguishes between the perception of bodies by humans and by God: 'The difference that exists between the appearances of bodies for us and their appearance to God is *quodammodo* that which exists between scenography and iconography.'[38] While in the first objects are represented using perspectival methods to create a three-dimensional depiction, the second depicts objects on a geometrical, horizontal level. Human perception is thereby described as inherently lacking in its representational capacities, missing the perspectival abilities of representing objects three-dimensionally, in comparison to God's perfect perception. In this case perspective is viewed as enabling the perception of objects in a manner that results from knowledge of their true nature.

In the *Theodicy*, Leibniz rather associates 'devices of perspective' with artifice and describes them as leading to error, in a way that is similar to his use of the term 'image' discussed before:

> Man finds himself the worse for this, in proportion to his fault; but God, by a wonderful art, turns all the errors of these little worlds to the greater adornment of his great world. It is as in those devices of perspective, where certain beautiful designs look like mere confusion until one restores them to the right angle of vision or one views them by means of a certain glass or mirror. It is by placing and using them properly that one makes them serve as adornment for a room.[39]

The two-faced 'devices of perspective' Leibniz describes seem like 'mere confusion' until viewed correctly, in the right angle or aided by a mirror or glass. As is often the case for Leibniz, the mirror offers a truer vision of reality. When viewed appropriately, the devices fill their function of adornment. Perception is mediated by the artful device that plays upon reflections, creating a shift in the observer's point of view. The mediation or intervention in perception through devices or apparatuses is discussed by Benjamin and Deleuze, specifically in relation to photographic and cinematic devices that bring about a shift in the viewer's perspective; it is a theme in their work to which I shall return.

iii Living mirrors

Although Leibniz does not describe the monad's perspective as located in space, but rather as a point in the scale between clear and distinct perceptions, on the one hand, and obscure and indistinct ones, on the other, the term 'point of view' relates to vision and projection, as manifest in his recurring descriptions of the individual as 'mirroring' the world.[40]

During the Baroque, researchers have shown, the concept of infinity was often represented by the metaphor of the mirror.[41] Leibniz begins using the image of a '*living* mirror', in a unique twist, rather late in his career. The description of the mirror as living may have been related to developments in his concepts of 'life' and 'living beings', which in later texts describe all substances as interconnected living beings, linked through the principle of pre-established harmony.[42]

The mirror, Leibniz writes, 'gives us a figurative expression, one suitable enough and already used by philosophers and theologians when they speak of an infinitely more perfect mirror, that is the *mirror of divinity*, which they made into the object of beatific vision'.[43] The 'living mirror' can therefore be understood as connecting these two elements: constant reflection of the deity by each substance and its interconnection with every other substance belonging to the infinite network that makes up the universe. The latter aspect is highlighted in Leibniz's better-known usages of the concept in 'Principles of Nature and Grace', for example, where he writes that 'each monad is a living mirror, or a mirror endowed with inner activity, representative of the universe, according to its point of view'.[44]

The 'living mirror' first appeared in a commentary by Leibniz on fragment 22 from Pascal's *Pensées*, composed around 1696 and later titled 'Double infinité chez Pascal et Monade'.[45] Leibniz uses therein the image of the 'living mirror' to describe his concept of the complete substance in contrast with Pascal's 'mite' [*ciron*]. In his interpretation of Leibniz's commentary, Buzon emphasizes that Leibniz's critique of Pascal hinges upon the latter's concept of the infinite extending in space alone. For Leibniz infinity is rather temporal, as the 'living mirror' image demonstrates.[46]

The 'living mirror' – an active, temporal, ever-changing representation of God and the universe – is comprised then of a kind of moving image. In a sense Leibniz's understanding of perception may be described as cinematic *avant la lettre*, which is no coincidence in a thinker for whom self-movement or force are the defining characteristics of individual substance.[47]

Leibniz's use of the 'living mirror' metaphor indicates that within his metaphysics, the universe and individual substances, as well as body and soul, are interdependent. Both relations are portrayed by Leibniz as instantiations of pre-established harmony.[48] The pervasiveness of round mirrors, usually concave or convex, during his times, suggests that Leibniz had round mirrors in mind when describing the monad as a 'living mirror', which corresponds to his description of God: a 'centre that is everywhere, but that his circumference is nowhere'.[49] The collective mirroring of all monads produces an image of God. There is thus a direct connection between individual and collective perception, and hence between what we may call individual and collective consciousness.

This means that even the minutest perception has an important role as mirror of the entire universe and God. Leibniz illustrates this by describing minds as 'architectonic samples or models [*échantillons*] of the Deity, capable of knowing the system of the universe', and elsewhere as 'machines'.[50] As discussed, he also uses the term 'detail' to describe the script, or plan that allows the monad to express the universe in a manner that harmonizes with the expressions of all other monads.[51] The significance of the notion of the detail to Benjamin may be related to his reading of Leibniz.[52]

2 Perception and body: Leibniz, Deleuze, Benjamin

i Having a body

Leibniz describes the monad's 'point of view' as belonging to its body. Rather than relating to its position in space, the point of view depends solely on the extent to which its perceptions, or the manner in which it expresses the universe, is clear and distinct. Since the monads are finite beings that express the infinite universe, as we recall, their expression is inevitably confused and indistinct to a certain extent. The vantage point of the monad is equivalent to its distinct zone of expression, as Deleuze writes in the section titled 'Having a Body', of *The Fold*. He quotes Leibniz as saying that that which he expresses clearly 'relates to my body'.[53] As Deleuze interprets Leibniz, this constitutes a moral deduction of the body, according to which the monad has a body because of its zone of expression, that is, because of its constituting a 'living mirror' that perceives the world: 'it is because we have a clear zone that we must have a body charged with traveling through it or exploring it, from birth to death.'[54]

According to Leibniz's understanding of the individual, the distinct point of view of each monad consists of the events that it will undergo, in an ideal, incorporeal form. In Deleuze's interpretation, this means that the point of view of the monad, its distinct zone of expression, consists not only of its body *per se* but also of what will happen to its body. The 'complete notion' of Caesar thus expressed clearly the historical event of the crossing of the Rubicon before it had taken place, pre-empting the contact of Caesar's body with its waters. As Deleuze writes, 'the monad expresses the world "according" to its body, according to the organs of its body, according to actions of other bodies upon itself.'[55] The point of view of the monad, its unique perspective, is embodied, and vice versa: its perspective, its zone of expression, constitutes its body. Moreover, since the perception of the monad includes that which will happen to its body, Leibniz's theory of perception is also a theory of the event.

Defined by its perceptions, it follows that the monad's self-movement corresponds to the flux in its perception, the becoming-distinct or becoming-obscure of its expression of the universe. In other words, the changes undergone by the monad are its movement from indistinct perceptions to distinct ones, motivated, as we have seen, by its appetition.

Deleuze highlights the significance of the obscure, dark part of the soul that necessitates the distinct, illuminated zone of expression that comprises the body. 'Contrary to Descartes, Leibniz begins in darkness. Clarity emerges from obscurity by way of a genetic process.'[56] As Deleuze explains, each monad is singular thanks to the obscurity of its perceptions, which are never completely distinct since the monad is finite and the expressed world infinite. The world only exists within the monads, and only in the form of 'representatives' or 'infinitely small actual elements', which Leibniz, as we have seen, also calls 'minute perceptions' and which Deleuze terms 'hallucinatory micro-perceptions'. These minute, confused perceptions are units composing our conscious, clear perceptions, as well as the passage from one perception to the next.[57] When we are in a state of sleep or unconsciousness, our minute perceptions are still active. Just as the body is never without movement, so the soul is never without perception, as Leibniz writes in *New Essays*.[58]

Within Leibniz's theory of minute perceptions, Deleuze underlines the constitution of perception mathematically, out of differential relations (as mentioned in Chapter 2). The relation between conscious and unconscious perceptions is not one of part and whole, Deleuze claims, but rather one of ordinary and remarkable, the singular and the multiple. Differential relations that constitute a singularity produce a conscious perception. By way of Maimon, Deleuze calls for a 'method of internal subjective genesis' that may replace Kant's method of conditioning, determining a concept of internal difference.[59] 'The physical object and mathematical space', writes Deleuze,

> both refer to a transcendental (differential and genetic) psychology of perception. Space-time ceases to be a pure given in order to become the totality or the nexus of differential relations in the subject, and the object itself ceases to be an empirical given in order to become the product of these relations in conscious perception.[60]

In Deleuze's understanding of perception, object and subject cease to be 'givens' and are produced through differential relations, and space-time becomes the nexus of these relations. Lacking an object, perception is always 'hallucinatory'.

Deleuze reformulates Whitehead's distinctions between microscopic and macroscopic prehensions in relation to Leibniz's theory of perception.[61] Thus, he distinguishes between 'macroperceptions' (conscious, clear and distinct perceptions) and 'microperceptions' (unconscious minute perceptions, 'pricklings of anxiety' [*aiguillons de l'inquiétude*] that render all perception unstable').[62] By underscoring the role of minute perceptions, which he calls the 'obscure dust of the world', Deleuze emphasizes the dark part of the soul, the disharmony and imbalance at the heart of Leibniz's philosophy and the Baroque era more generally.[63] Consciousness and distinct perception necessitate unconsciousness and indistinct, foggy ideas.

Leibniz's emphasis on unconscious states, as well as those of dizziness or hallucination, is foregrounded against Descartes' cogito and 'clear and distinct ideas', challenging Leibniz's undisputed reputation as an optimist. Deleuze clearly finds in Leibniz a reply to the philosophy of representation in the form of a theory of perspective linking body and event. Deleuze would have perhaps described Leibniz's monad, then, as a Claude glass, that is, a tinted *dark*, or partially darkened, mirror of the universe.[64]

ii If nature changes, the body's perception changes, too

The interrelation between the concept of 'perception' and the experience of the human body first appears in Benjamin's early texts, specifically in several fragments written at the end of the 1910s and early 1920s.[65] Several recent studies have called attention to how the theme of embodied experience elaborated in these fragments persists in his writing, taking form in his later work as the 'collective physis', depicting the collective as body.[66] In what follows, we interrogate the interdependence between individual and collective perception, and the notion of perception as differentiation, in conjunction with Benjamin's reading of Leibniz.

In 'Outline of the Psychophysical Problem' (written in 1922–3, but unpublished in his lifetime), Benjamin distinguishes between the 'lived body' [*Leib*] and 'thing body' [*Körper*], a distinction made by phenomenologists at the time.[67] The two corporeal modes are distinguished by the manner in which the body appears in our self-perception: while the 'lived body' appears to our inner eye, the 'thing body' denotes our external perception of our body. Benjamin distinguishes between 'lived experience' [*Erlebnis*] and 'extended experience' [*Erfahrung*], understanding corporeal experience primarily through perception.

In the fragment 'Perception and Lived Body' [*Wahrnehmung und Leib*] Benjamin writes that 'through our corporeality [*Leiblichkeit*], ultimately most immediately through our own body, we are placed in the world of perception [*Wahrnehmungswelt*]'.[68] The manner in which we perceive our body shapes our external perceptions, forming our experienced world through a 'hierarchy of distances', which may even lead to a kind of estrangement from our body, deriving from the fact that there are many parts of the body that we do not usually see.

While, in 'Outline of the Psychophysical Problem', Benjamin speaks of mind and body [*Geist und Leib*] as unified, he mentions, in the same breath, a rupture between 'lived body' and 'thing body'. 'With the "lived body", man belongs to mankind, with the "thing body", to God', Benjamin writes. 'Both have fluctuating boundaries with nature, the expansion of both affects the world's events, and for the profoundest reasons', he continues.[69] 'Lived body' and 'thing body' are continuous with nature. Perception links these two oppositional aspects of the body and the world. In the following, Benjamin defines perception as a 'differentiation':

> The representation of total vitality in life causes fate to end in madness. For all the living reactivity is bound to differentiation, whose permanent instrument is the body [*Körper*]. This is an essential function of the body. Only the body can be viewed as an instrument with which to differentiate between the vital reactions and at the same time can be comprehended in terms of its psychic animation. All psychic energy can be differentially located in it, just as the old anthroposophy attempted this when it regarded the body as an analogue of the macrocosm. One of the body's most important organs of differentiation is perception. The zone of perception also shows clearly the variability to which it is subject as a function of nature. If nature changes, the body's perception changes, too.[70]

The body is defined through differentiation that is performed through perception, in a formulation that is very close to Leibniz's understanding of perception as constituted by differential relations, and as constitutional of the body. Benjamin uses terminology that relates to forces, describing the body as acting and reacted upon by 'vital animations' and containing 'psychic energy'. Perception is described as a uniquely corporeal function. The body's 'zone of perception' defines its relation to the 'macrocosm' or exterior world that it continually expresses, changes in nature thus leaving their mark on the body's perceptions, as in Leibniz's pre-established harmony.

As is also evident in the above citation, in these early 'anthropological' sketches Benjamin pays special attention to extreme states of perception, such as those incurred by states of madness. In a fragment (dated to 1917 or earlier) titled 'Perception Is Reading', Benjamin links perception and madness: 'In perception [*Wahrnehmung*] the useful (the good) is true [*Wahr*]. Pragmatism. Madness is a form of perception alien to the community. The accusations of madness levelled against great reformers. Inability of the masses to distinguish between knowledge and perception', reads the fragment.[71] Madness is described as a 'form of perception' which society does not comprehend, leading to the levelling of (false) accusations of madness at misunderstood visionaries.

iii Body and nature

In the Second Version of the Artwork essay, written years later in 1936, we find once again a conjunction of body and perception. As we recall, this seminal essay defines a shift in the nature and experience of art brought about by its 'reproducibility' that entailed a loss of the artwork's 'aura'. Benjamin relates this shift to the human body, suggesting that the possibility of splitting and splicing together an actor's performance when shooting a film indicates that art no longer belongs to the realm of 'beautiful semblance'.[72] In other words, the distinction between truth and semblance, and the artwork that was grounded upon this distinction, necessitated a wholeness of the body.[73] Benjamin compares the self-alienation produced by representation of the actor's body by the cinematic apparatus to that produced by one's mirror-image – a recurrent theme in the writing of the Romantics. The difference between mirror-image and film-image is that the latter is detachable, tearing the semblance [*Erscheinung*] of the body from its 'origin' and relocating it to 'a site in front of the masses'.[74] The possibility of splitting and reassembling the body defines, then, the 'age of the assembled [*montierbar*] artwork'.[75] Here, Benjamin anticipates concerns that have only grown in contemporary times, relating to changes in human perception due to the lack of corporeal presence in the virtual world.[76]

For Benjamin, mimetic representation is primarily that of the body, and the biblical prohibition against representation – 'Thou shall not make unto thee any graven image' – is a moral injunction that relates specifically to the implications of producing images of the body. This is an argument he makes in *Origin of the German Trauerspiel* and repeats in the Artwork essay. In the latter, he suggests that the shift in the collective mode of perception was brought about by alienation from the self, caused by the splitting of the body and its fragmentation by newly developed technologies.[77] These ideas are echoed in a comment he makes in a footnote tracing the origin of mimesis to the body of the mime.[78]

The parallel Benjamin draws in the Artwork essay between the experience of art and the experience of nature has not received much attention, and has perhaps been predetermined by Adorno's early reading.[79] Benjamin refers therein to transformations in the relationship between humans and nature brought about by technological developments through a distinction between 'first' and 'second' technologies.[80] 'The first technology really sought to master nature', he writes, 'whereas the second aims at an interplay between nature and humanity'.[81] The function of the second technology, the chief example of which is film, is to train humans in adapting to the technological apparatus that has gained increasing control over their lives.[82] Benjamin's 'first technology' refers to early modern technologies such as optical inventions like the telescope and microscope. 'Second technology' seems to indicate a second shift in the relation to nature that recognizes that nature may not be completely controllable by technology, rather both humans and nature are at the mercy of the technologies they developed.

In the section 'To the Planetarium' from *One-Way Street*, Benjamin also refers to the process of alienation between nature and humanity, which is echoed in a splitting or fragmentation of the body itself. Both are incurred by optical developments, suggesting a rupture between the visual, on the one hand, and the tactile and sensual, on the other, that took place in early modernity. This process of distancing created an estrangement from nature, which in turn led to its exploitation by humans, he suggests. The continuity between humans and nature, which Benjamin also calls a 'community' [*Gemeinschaft*] with nature, was replaced by a cataclysm or rift. Thus, community with nature was substituted by a society [*Gesellschaft*] of humans, or, as Benjamin refers to it in the text, by *man*kind [*Menschheit*].[83]

Leibniz is not the target of Benjamin's implied critique of early modern science, which remains focused on astronomers Copernicus, Kepler and Brahe as well as Descartes.[84] While Leibniz took part in the optical revolution, his philosophy arguably predates the modern break from nature, the origins of which many ascribe to Descartes' mind-body dualism. As Phemister shows, Leibniz's theory of expression implies an interconnectedness between all monads, whether they belong to humans, animals or inanimate beings.[85] As we have seen, body and soul are intimately connected, as well, through Leibniz's theory of perspective.

3 Benjamin's historical perspectivism

i Storytelling and montage

In his concepts of 'image' and 'montage', Benjamin drew on the long philosophical tradition linking perception and visuality, and, to some extent, specifically on Leibniz's perspectivism as constitutive of individuality. The concept of 'montage' [*Montage*] becomes increasingly important in Benjamin's work from the 1930s. To some extent, this relates to his exposure to cinema and cinematic montage. However, cinematic montage is only one form of montage out of many that Benjamin discusses; alongside literary montage, photomontage, the montage of objects within Dada artworks and historical montage, which serves as my focus here.[86] The German *Montage*, from the French *montage*, designates mounting or setting up, bringing together or connecting. As we have seen, Benjamin uses the term not only in relation to art but also in relation to history and historiography.

In 'Crisis of the Novel', Benjamin describes montage as an epic form – the key stylistic principle [*Stylprinzip*] of Alexander Döblin's novel *Berlin Alexanderplatz*.[87] Montage 'explodes the framework of the novel … and clears the way for new, epic possibilities', writes Benjamin. Yet despite the use of this explosive stylistic principle, Döblin's novel remains an 'old bourgeois *Bildungsroman*' – although it pushes this genre to its extreme. Montage, in this case, is still at the service of narrative.

In Benjamin's writings from the late 1930s on, the destructive force of montage intensifies, destroying not only the framework of the novel but the epic as well; or, put differently, montage becomes a counter-narrative principle. As Benjamin writes in a note while composing 'On the Concept of History': 'The second fortified position of historicism is the idea that history is something that lets itself be narrated [*das Sich-erzählen-lassen*]. In a materialist investigation, the epic moment will always be blown apart in the process of construction.'[88] In historical-materialist investigations, the destruction of epic narrative constitutes part of the construction, or montage, of historical images.

Just like the 'historical object', the 'dialectical image' is described in the *Arcades Project* as a monad. As Benjamin writes, 'the object constructed in the materialist presentation of history is itself the dialectical image'.[89] This is somewhat confusing since the dialectical image is also described as a kind of internal, mental image that flashes up when thought stands still, while the 'historical object' denotes an external object of analysis.[90] It is precisely the monadological structure which allows this tension between exterior and interior, creating continuity between 'the knower and the known'.[91] As in Leibniz's case, the conception of the dialectical image therefore hinges on a conception of perception that does not distinguish between the analysing subject and the object of analysis. Benjamin does not mention a subject, yet he does mention 'thought' and 'consciousness'. At the same time, Benjamin writes that

> if the object of history is to be blasted out [*herausgesprengt*] of the continuum of historical succession, it is because its monadological structure demands it. This structure first comes to light in the extracted object itself. And it does so in the form of the historical confrontation that makes up the interior (and, as it were, the bowels) of the historical object, and into which all the forces [*Kräfte*] and interests of history enter on a reduced scale [*verjüngtem Maßstab*].[92]

The structure of the object of history as monad demands, on the one hand, that the object be separated, extracted out of the continuum of history, thereby enabling the construction of a non-continuous historiography. Benjamin contests the notion of the historical continuum that describes history as teleological and progressive. On the other hand, this same monadological structure enables the object of history or dialectical image to enclose within its form the conflicting forces and interests of history through the tapering of their scale. Thus while he contests the notion of history as continuous and progress-driven, Benjamin describes the notion of history as motivated by forceful, conflictual powers. The image is created by a break in movement, yet it contains dynamic forces, necessary as well for the construction of history. The concept of the 'dialectical image' encapsulates, once more, the tensions between continuity and discontinuity, construction and destruction. By highlighting destruction alongside construction, Benjamin's distances himself in his late use of the term 'monad' from Leibniz's description of the monad as striving towards perfection. Yet he retains Leibniz's conception of the monad as independent and self-motivated by an inner force.

ii Montage as procedure and as principle

Benjamin's concepts of dialectical image and montage, we will see in this chapter and the next, include a historical version of perspectivism, informed by Leibniz. As Buck-Morss suggests, the 'dialectical image' in itself enacts a kind of montage, through its dialectical construction.[93] While Benjamin describes montage, in the Artwork essay, as a film-making procedure mediated by the apparatus, some of his other references to montage denote it as a procedure that merely exhibits or shows distinct elements of material or images, allowing meaning to emerge from their juxtaposition.[94]

Described as an artistic *procedure* [*Verfahren*] in 'The Author as Producer', montage functions as a *philosophical principle* [*Prinzip*] in 'Convolut N' of the *Arcades Project* and 'On the Concept of History'.[95] The term thus oscillates in Benjamin's different works, between two apparently oppositional meanings. On the one hand, the *procedure* of montage 'The Author as Producer' focuses on montage as a theatrical *method* which interrupts action. On the other, the *principle* of montage in the *Arcades Project* describes montage as a philosophical principle of construction [*Konstruktion*].[96] Both as procedure and as principle, montage may be defined as a dual-staged process including either interruption then arrangement or destruction and construction. The principle of montage, we will see, bears a stronger link to Leibniz than the procedure of montage.[97]

In 'Convolut N' of the *Arcades Project*, Benjamin describes montage as a principle that produces a clarity or perceptibility of history, the possibility of watching history as one watches moving-images, as well as a principle of construction serving historical materialism:

> Must the Marxist understanding of history necessarily be acquired at the expense of the perceptibility [*Anschaulichkeit*] of history? Or: in what way is it possible to conjoin a heightened graphicness [*Anschaulichkeit*] to the realization of the Marxist method? The first stage in this undertaking will be to carry over the principle of montage into history. That is to assemble large scale constructions out of the smallest and most precisely cut components. Indeed, to discover in the analysis of the small individual moment the crystal of the total event.[98]

Here, montage is described as a principle [*Prinzip*] which can bring together 'graphicability' [*Anschaulichkeit*], on the one hand, and the historical materialist approach to history, on the other. Its method is described as that of construction – using parts that are constructions, or assemblies of even smaller parts [*konfektionierten Baugliedern*], themselves. The principle of montage allows the (Marxist) historian/philosopher to follow the convictions that truth is contingent upon historical context rather than timeless, and ephemeral objects no less worthy of philosophizing than eternal ones. In montage the small, 'precisely cut' components are paramount for the construction of the whole. Moreover, by exercising this principle, totalities emerge through analysis of minute details. Montage is described not only as a principle for the conjoining of disparate material, or sections, but also for analysis of 'moments' or time segments, enabling the discovery of a 'total event', that is, a whole or totality contained by each segment, just as the monad contains the universe. In other words, montage is defined as a monadological principle.

Benjamin does not describe the process of montage as involving mediation of the assembled material. As he writes, the historical materialist must 'discover' rather than construct the 'total event'. Adorno criticized Benjamin's 'montage', in an essay published in his 1955 book *Prisms*, for its lack of mediation in the form of commentary:

> Benjamin's intention was to eliminate all overt commentary and to have the meanings emerge solely through a shocking montage of the material. His aim was not merely for philosophy to catch up to surrealism, but for it to become surrealistic ... The culmination of his anti-subjectivism, his major work was to consist solely of citations. Only seldom are there interpretations noted which could not be integrated into the Baudelaire study or the theses 'On the Concept of History', and there is no canon to indicate how the audacious venture of a philosophy purified of argument might be carried out, or even how the citations might be meaningfully ordered. His philosophy of fragmentation remained itself fragmentary, the victim, perhaps, of a method, the feasibility of which in the medium of thought must remain an open question.[99]

Benjamin uses montage in the composition of the *Arcades Project* as an arbitrary process, in which material is assembled and ordered randomly, Adorno contends. Benjamin's philosophy, he writes, becomes 'surrealistic'. Surrealist practices invited chance to intervene in the process of creation, as poet André Breton writes: 'It is even permissible to entitle POEM what we get from the most random assemblage possible (observe, if you will, the syntax) of headlines and scraps of headlines cut out of the newspapers.'[100] Adorno's criticism implies Benjamin's 'shocking montage' of material in the *Arcades Project* is comparable to such poems. He views Benjamin's excessive use of quotations therein as dangerously pushing to the extreme his statement, from *Origin of the German Trauerspiel* according to which 'truth is the death of intention'.[101]

In his grand-scale project, Adorno suggests, Benjamin was 'carried away' so to speak, by his use of citations, until the citations outnumbered the commentary. While in earlier works, such as *One-Way Street* and *Origin of the German Trauerspiel*, Benjamin integrated the extensive citations within a main text, in the *Arcades Project*, citations seemed to have taken priority, some of the convolutes almost lacking mediating commentary. Adorno casts doubt upon whether this 'philosophy of fragmentation' may function as philosophy, implying that philosophical argumentation necessitates a subjective intention and point of view.

Yet, as suggested, Benjamin defines montage as a monadological principle, and as such, it is not arbitrary as the random assembly of newspaper cut-outs on a page. It involves the creation of meaning through the juxtaposition between parts, as time and space are created relationally in Leibniz's philosophy, or as in the renowned 'Kuleshov effect', meaning emerges from the juxtaposition between successive shots.[102] Thus in the case of the *Arcades Project*, meaning is created relationally, by the constellation of fragments collected in every 'convolute'. The reader, entrusted with the decision of the order in which to read the fragments, becomes an active participator in the construction (and destruction!) of meaning.

Although it is impossible to know whether Benjamin viewed the *Passagen* as a blueprint to a future project or as a finished work, I believe, with Eiland and McLaughlin, that 'the research project had become an end to itself'.[103] That is, Benjamin had probably not intended to add commentary to the fragments at a later stage, viewing montage as a methodological principle for the construction of this work.

His preoccupation by the question of the extent to which it is necessary to mediate historical materials is manifest in 'The Storyteller', wherein Benjamin compares the historian to the chronicler. While the historian is tasked with explaining events, the chronicler interprets events by the way they are 'embedded in the great inscrutable course of the world'.[104] In chronicles, events are displayed 'as models' in relation to the totality of events, while the relation between different events within this totality is unexplored. In the model of historical construction Benjamin proposes, connections between events are determined, yet not by relations of cause and effect.[105]

By developing the principle of montage, Adorno seems to suggest that Benjamin incorporates, with questionable success, a tradition associated with the essay and shorter-form writing into his longer works. In 'The Essay as Form' (1958), Adorno describes Benjamin's writing as embodying the archetypal form of the essay. Adorno evokes Leibniz, drawing a link between Leibniz's monad, the Romantic fragment and the modern essay, yet at the same breath also denying the debt the modern essay owes to Leibniz, implying Leibniz had not taken intellectual freedom far enough.[106] Adorno describes the essay as constructing 'immanent criticism of cultural artefacts', while 'the artefact is a monad, yet it is not'.[107] The artefact cannot quite be a monad because of the totality included within the monad. The Romantic fragment, Adorno seems to suggest, is a more fitting image of the artefact, in its incompletion, resisting totality through infinite self-reflection.

My reading, however, following Benjamin and Deleuze, suggests that in Leibniz's writing there is already a resistance of totality and perfection, alongside the movement towards perfection, a resistance to which Benjamin was perhaps more attuned than Adorno, and which Deleuze explicitly highlights in his reading of Leibniz in *The Fold*. This movement away from perfection and towards destruction manifests itself in the aspects of discontinuity of Benjamin's principle of montage.

Buck-Morss suggests that a key concern for Benjamin in the Artwork essay – the way in which technology affects the human sensorium – harkens back to the original meaning of the term 'aesthetics', and that art is thereby redefined, no longer distinguishing itself by its separation from 'reality'.[108] As she reminds her readers, 'aesthetics' is derived from the Greek *Aisthesis*, meaning the sensory experience of perception. She follows Eagleton, who traces the reversal in the meaning of the term that resulted in its being applied primarily to art during Benjamin's time.[109]

Buck-Morss does not problematize the conception of 'image' in this essay.[110] Just as Benjamin understands art and aesthetics through sense perception, sensory experience is also a key to Benjamin's shifting conception of the image. Following Buck-Morss, we can understand the transitions in Benjamin's conceptualization of the image as resulting from his diagnosis of a crisis in perception and experience discussed previously. In the Artwork essay, as Buck-Morss describes, technology is seen as both the central cause of 'perceptual splitting' during the modern era and its possible cure.[111]

In earlier works, from the 1917 fragment 'Perception as Writing' through to the Surrealism essay (1929), extreme perceptual experiences such as intoxication, madness and experiences of the unconscious lead the way to a reversal of the crisis through the dissolution of the bourgeois self. Moreover, the understanding of individual perception as linked to collective perception, and the collective as body – underscored by Buck-Morss in the 'Artwork' essay – is already present in Benjamin's earlier texts.

His understanding therein of the relation between individual perception and collective bodies, as discussed, echoes Leibniz's theory of perception, in which individual perception expresses the universe and God's 'perception'. Leibniz's theory of an unconscious composed of minute perceptions filtered into consciousness by differential relations was also influential for Deleuze's conceptions of the image. In Leibniz, as we will see, the image is linked not only to body but also to time – and to the body *in* time.

4 Deleuze's mannerist perspectivism

i Leibniz's perspectivism

Deleuze describes Leibniz's thought as a 'profound mannerism'.[112] Alongside the Stoics and Whitehead, and against Descartes and Aristotle, Leibniz is considered by Deleuze as a mannerist thinker, and his mannerism takes specific form in his perspectivism, or theory of points of view. 'Point of view on a variation replaces the centre of the figure or a configuration.'[113] Individuality is constructed and defined through constant variation rather than stable qualities. Deleuze views Leibniz's perspectivist conception of individuality as an alternative to Kant's understanding of experience through subject and object. In one of his lectures, Deleuze describes Leibniz's perspectivism in the following way:

> Leibniz cannot stop. He has to go all the way to a theory of point of view such that the subject is constituted by the point of view and not the point of view constituted by the subject … The individual notion is the point of view under which the individual expresses the world. It's beautiful and it's even poetic … For Leibniz, every individual substance is like an entire world and like a mirror of God or of the whole universe that each substance expresses in its own way: kind of like an entire city is diversely represented depending on the different situations of the one who looks at it.[114]

Deleuze recognizes the originality of Leibniz's designation of the individual through its point of view, rather than vice versa. As he writes, Leibniz was the inventor of a 'theory of point of view' that was the precursor of perspectivist theories put forward by a diverse array of authors, namely Nietzsche, William and Henry James, and Whitehead.[115] In all of the above, perspectivism entails a kind of relativism, Deleuze writes, yet not the kind of relativism we suspect, according to which truth changes according to the subject. Rather, what changes are 'the conditions under which the truth of a variation appears to the subject'.[116] In Serres' words, which Deleuze quotes, Leibniz's perspectivism is viewed as 'the truth of the relative rather than the relativity of truth'.[117]

Perspectivism – whether that of Leibniz, Whitehead or Nietzsche – views the world as reducible to the multiple points of view of individuals upon it. Thus, for Whitehead, 'apart from the experiences of subjects there is nothing'.[118] Similarly, for Leibniz as interpreted by Deleuze, the world itself consists of a virtuality included in every soul which expresses it.[119] The world is folded within the expressing souls, and as such exists in actuality only in its being enveloped by them. Deleuze follows Serres in understanding Leibniz's perspectivism as a symptom of his 'decentred' age: the multiple points of view of the monads replace the lost centre.[120] While perspectivism consists of a pluralism, in the sense that it implies plural, separate points of view, in Deleuze's reading this does not entail discontinuity between these points of view. Deleuze argues, rather, that they are distant from one another yet continuous.[121]

ii Leibniz as a mannerist thinker

The distinction between Mannerism and Baroque has been the subject of many art-historical debates. Wölfflin, and other nineteenth- and early twentieth-century critics, did not recognize Mannerism as a distinct period. Yet in the 1920s and 1930s, critics like Walter Friedlaender, Max Dvorak and Richard Sayce began to recognize Mannerism for its originality and innovation, distinguishing more clearly between High Renaissance, Mannerism and Baroque.[122]

Mannerism was thus only theorized as a period in art history in the twentieth century, often through its comparison with expressionist and abstract art.[123] It was described as a reaction to High Renaissance, an 'anticlassical' style, in which artists no longer strove towards an idealized depiction of objects, rather focusing on intellectual and technical virtuosity.[124] The paintings of El Greco, depicting elongated figures with disproportionally small heads, have become a paradigmatic example of the Mannerist style.

Deleuze describes an 'operational relation' between Mannerism and the Baroque.[125] By doing so, he contests the inferior status of Mannerism as a degeneration of classicism, as well as the very separation between the two periods. Like other post-war art critics, Deleuze compares Mannerism and Baroque to modern and Abstract art, suggesting a shared excessive impulse to fill the canvas with folds.[126]

El Greco's paintings serve, for Deleuze, as a main example for this mannerist driving principle, by introducing infinite folds onto his canvases. As he writes of El Greco's *Christ in the Olive Garden*, in which the folds in Christ's tunic, and in the blankets and clothing of the sleeping disciples in the forefront of the painting, are echoed in the rock in the background, 'folds seem to be rid of their support – cloth, granite or cloud – in order to enter into an infinite convergence'.[127] The folds are no longer there merely to cloak the finite body but also express the infinity of nature.[128]

Like the Baroque, Deleuze views Mannerism as extending beyond its definition as an art-historical period. It is, at the same time, a philosophical categorization to which the Stoics, for example, belong as well, and which operates according to what he calls the 'logic of the event'.[129] As discussed in Chapter 1, Leibniz's individual is defined according to his or her predicates – or 'events', as Deleuze calls them – rather than by essential qualities. Here, once again, Deleuze contrasts Leibniz's Mannerism with Descartes' essentialism: while the latter defines substance according to its essential attributes, Leibniz's individual is fluidly constituted through spontaneous 'modulations'.[130]

Not only the mannerist subject but also the object is similarly defined not as the sum of its parts but by these parts' inseparability, which results from the ways in which they are folded. Mannerism thus constructs its objects through qualitative relations. Deleuze names the 'new object', defined through its function rather than essence, an *objectile*. The *objectile*, according to Deleuze, exists only through its morphing nature, with no stable object or substance enduring through its changes.[131]

The inspiration for the *objectile* came from architect and theorist Bernard Cache's concept of 'non-standard production' developed in *Earth Moves*, (which later led to the development of 'non-standard architecture').[132] The intellectual exchange with Cache, who attended Deleuze's seminars, seemed to have played a significant role for *The Fold*.[133] Cache suggests that the object must be redefined, following new production processes using Computer-Assisted Conception and Fabrication (CFAO) systems, through which unique objects can be produced industrially. We have moved beyond the age of technical reproducibility based on the repetition of a model, 'from the mold we move towards modulation', as he writes.[134] Cache interrogated the revolution in architecture following the introduction of digital technologies into its production processes, laying the ground for what has recently been described as a 'digital turn' in architecture.[135] When considering Leibniz's own contribution to the development of numerical language, the convergence between his ideas and those of Cache is less surprising.[136] Building on Cache, Deleuze describes the new object or *objectile* as one that 'no longer refers its condition to a spatial mold – in other words, to the relation of form and matter – but to a temporal modulation that implies as such the beginning of a continuous variation of matter as a continuous development of form'.[137]

Deleuze brought Cache's concept of 'image inflection' and Leibniz's 'point of view' together as well.[138] For Deleuze, Leibniz's 'point of view' is the condition for the subject and object's variation. It is the site on which the curve is inflected, or the variation occurs. Deleuze understands inflection as the 'pure event' and the *objectile* as the process of becoming an event.[139] In defining these processes of creating inflections, or folding, he foregrounds, like Benjamin, the temporal aspect of Leibniz's conception of individuality.

When writing that Leibniz's thinking is mannerist, Deleuze means, then, first, that his conception of perspectivism, through which individuality is defined, is non-essentialist and modulated. In today's terms, we could even say algorithmic. Yet he highlights the 'operational' function of Mannerism in relation to Baroque for another reason as well. As he puts it, the second aspect of Mannerism is 'the omnipresence of the dark depth which is opposed to the clarity of the form, and without which manners would have no place to surge forth from'.[140] Leibniz's thought is defined as Mannerist, then, also on account of its complex, sombre qualities, informed by novel developments in optics at the time, that understood clarity as inseparable from obscurity.[141]

Deleuze gives prominence to the developments in the sciences of light and colour during the Baroque and their relation to new techniques in painting such as the use of a dark red-brown background against which the lighter forms emerge, in the works of Caravaggio and Tintoretto.[142] Following Wölfflin, he describes the techniques of chiaroscuro as a progressivity of light that is augmented or reduced by degrees; what appears to be lit becomes so only in relation to areas of darkness. 'Clarity ceaselessly plunges into obscurity', as Deleuze writes.[143] He refers to Leibniz's 'theory of shading', and the manner in which Leibniz's conception of gradated intensivity informed the use of light and colour by mannerist and Baroque painters. This theory reappears in Deleuze's later discussion of Orson Welles' cinema, making manifest the link between Deleuze's conceptions of Mannerism and Modernism.[144]

Finally, in defining Leibniz as a mannerist, Deleuze highlights the fluidity and transformability of his concept of individuality and his metaphysics more generally. He describes Leibniz's principles, as well, as interchangeable, to some extent, or folding into each other. He thereby underlines the excess of Leibniz's philosophy, and of the Baroque, in generating principles, manners and folds that cannot be contained.[145]

Deleuze notes these elements of the Baroque in his description of Mannerism as a trait of schizophrenia, both in *The Fold* and in *Cinema 2*. In this context, Mannerism is described as creating a different point of view of the world, bringing things closer or creating distance.[146] Leibniz's best of all possible worlds becomes, in Deleuze's reading, a mannerist, over-perfect and therefore distorted vision that converges with Deleuze's critique of the image of thought as expressed in the conception of the distorted image or simulacrum.

Benjamin does not discuss directly the relationship between Mannerism and Baroque, but he does mention mannerist painting, and specifically El Greco, in passing:

> Just as the painting of the mannerists knows nothing of composition in placid light, so the theatrical figures of the epoch stand in the glare of their wavering resolution. In them is expressed not so much the sovereignty that their stoic manner of speech brings to the fore as the sudden caprices of a continually changing storm of emotions, in which the figures of Lohenstein in particular emerge like torn, fluttering flags. Not entirely dissimilar are the figures of El Greco, in the relative smallness of their heads, if it is permissible to understand this expression in a figurative sense. For it is not thoughts but fluctuating physical impulses that determine them.[147]

Benjamin compares, in this passage, the shifting lighting in mannerist painting with the indecisiveness of the characters in Trauerspiel plays, especially those by Lohenstein. These characters, in their fluctuating positions and actions, are moved by impulses rather than thoughts.[148] Benjamin thus underscores the corporeal aspect of Baroque individuals, in which subjects become more object-like. Here too, Baroque drama is presented as predating the optical revolution and modern-era conceptions of reason that severed the continuity between body and mind.

5

From the crystal palace to cinematic crystals: Baroque optics as pre-cinematic form

Introduction: Image and counter-image

This final chapter focuses on the different ways in which Baroque allegory and Leibniz's perspectivism function as precursors of cinematic form, according to Benjamin and Deleuze, and on the inspiration they find in the Baroque for their modern cinematic and photographic theories. In the work of both thinkers, we find image theories that consist of critiques of the hegemony of visual and representational thinking. Benjamin and Deleuze reconfigure subject-object relations in a manner that promotes a non-hierarchical, de-centred, 'real rather than possible' (in Deleuze's terms) conception of experience.

As we have seen in the previous chapter, the concepts of 'image' used by Deleuze and Benjamin are temporal, embodied and rooted in sense perception. Both draw on a philosophical tradition in which the 'image' is linked with memory and imagination, yet their conceptions of image relate to not only the mental but also corporeal image. Their conceptions of 'image' call into question mimetic representation. By emphasizing sense perception, they challenge the age-old philosophical link between truth and vision.

The foundational role of Leibniz's perspectivism for modern cinema and literature is embodied in the myth of the crystal pyramid that seals the *Theodicy* and attempts to justify the mutual existence of divine justice and human freedom. The myth, which suggests the co-existence of many virtual worlds, informs Benjamin and Deleuze's understanding of virtuality as well as their key image-concepts, namely the 'dialectical image' and the 'crystal-image'. These concepts are similar in that they bring together image and ruptured temporality, yet they also render the divergences between Deleuze and Benjamin's thought most evident. While for Benjamin the dialectical image may enable a true perceptibility of history, for Deleuze, the crystal-image is a falsifying mechanism.

In their conceptions of the image, the divergences between Deleuze and Benjamin's thought are also apparent. The etymological distinctions between *image* and *Bild* must be taken into account. *Image* originates from the Latin *imagio* that relates to *imitate*, hence the term 'image' inherently designates a mimetic relation. The Germanic *Bild* comes from the verb *bilden*, equivalent to *gestalten*, forming, and relates also to *Bildung*, education. *Bild* therefore echoes formation, yet unlike *image* it does not necessarily imply the act of imitating or producing a copy. The German *Abbild*, designating image as well as reflection, or likeness, is closer in meaning to *image*.[1]

Throughout his writing, Deleuze engaged with the relations between the image and thought, the visible and the thinkable. As a form of representation, which, according to Deleuze, consists of distortion, 'image' does not necessarily belong to the realm of thought. In *Difference and Repetition* the concept of 'image of thought' is designated as pre-philosophical, consisting of commonsensical presumptions that are presupposed by philosophical thought. The concept of simulacrum, the distorted image of modernity, is perhaps an attempt to delineate an affirmative concept of the image, in this work, yet by designating the 'image of thought' as non-philosophical, Deleuze paradoxically adopts some of the traditional preconceptions of philosophy, assigning the image and visual field more generally to non-conceptual domains.

The intricate relations between thought and the image are a persistent theme in Benjamin's writing as well. Thought and image are conjoined in the 'thought/thinking image' [*Denkbild*], a concept Benjamin uses to describe the aphoristic texts assembled in *One-Way Street* and *Berlin Childhood around 1900*.[2] While Benjamin describes images as appearing in the mind, memory and dreams, he also describes the image as presented in language. Common to all image forms is the fact that the image may enable access to truths. Thus, for example, the dialectical presentation enabled by the image can disclose 'true historical time'. Deleuze and Benjamin's discussions of cinema and conceptions of image-montage, in which they draw on Leibniz's perspectivism, expose the similarities in their approaches as well as their divergent positions vis-à-vis the image.

1 Allegory, image, metaphor

i Allegory and symbol

The last section of Benjamin's *Origin of the German Trauerspiel* positions the allegory as a central formal technique of the Baroque mourning play. Allegorical composition usually refers to a text that contains a double meaning, one on the surface and one 'hidden beneath' it. The allegory has often been linked with the symbol, and the two terms, of Greek origin, were sometimes even considered synonymous.[3] Benjamin, however, differentiates between the allegory and the symbol, as part of his larger argument on the distinctions between Classicism and Baroque, and between ancient tragedy and modern *Trauerspiel*.

Against simplistic definitions of the allegory as the relation between a signifier and its signification, Benjamin views it as a philosophical concept.[4] Specifically, he defines the allegory as 'expression', a concept that, as we demonstrated in Chapter 1, relates to Leibniz's theory of expression. 'Allegory … is not a perfunctory image-technique [*spielerische Bildertechnik*], but expression [*Ausdruck*], as language is expression, as indeed writing is expression', he writes.[5] According to Leibniz's concept of expression, as discussed, the expresser and the expressed correspond to one another in their inner relations, or proportions. The similarity between them, that Leibniz describes as a congruence or proportional similarity, is internal, not external as in an image. This is true also of Benjamin's allegory and the ideas that it expresses, for he insists that that which is expressed are ideas, and not meanings or significations. While the allegory sometimes uses images or visual techniques, its form is closer to writing, in which the relation between the written version of words and their meaning is not necessarily visual, and rather relates to the laws and internal logic of the language.

Benjamin likens the allegory to a ruin – a fragment marked by history, which is a part of another whole and at the same time exists independently.[6] As we recall, in Leibniz's concept of expression, the whole is defined by the part as much as the parts are determined by the whole. To say that every monad expresses the totality of monads is equivalent to saying that each monad contains, in a sense, the entire universe. They are co-dependant; each is incomplete without the other. The same kind of interdependence can be said to exist between the ruin of an ancient building and the building in its entirety as it existed in the past, or in memory, or between an allegory and the idea it expresses. The allegory contains an abbreviated version of a larger idea, or a longer history, in nuce.

In creating a parallel between allegory and ruin, Benjamin mentions, specifically, classical Greek or Roman ruins, such as those sometimes represented in the set design in German mourning plays. He describes the allegory as an anti-classical form, in that it challenges the organic and perfect ideals of classicism. Contrary to these, it contains dialectical tensions and antinomies. In this sense, Benjamin's concept of allegory is comparable to Deleuze's concept of Mannerism (and to some extent the latter was informed by the former).

Benjamin moreover refers, in his description of Baroque allegorism, to the Baroque 'feeling for style', or Mannerism, which compelled Baroque authors to 'heap up fragments uninterruptedly, without any well-defined idea of a goal, and, in the unremitting expectation of a miracle, to introduce stereotypes for intensification'.[7] Baroque authors reused existing materials excessively and haphazardly, attempting to create new meaning by jumbling fragments together. As a construction whose artifice is visible, the allegory stands in stark contrast to the symbol, described by Benjamin as an 'image of organic totality'.[8] In the constructed nature of the allegory, as will be discussed, also lies its link to Benjamin's procedure of montage, as defined in the *Arcades Project*.

Benjamin quotes philosopher Joseph Göress, who describes the allegory as dynamically taking on 'the flux of time itself'.[9] Unlike the momentary, fleeting temporality of the symbol, the allegory has a duration; it unfolds in time. This aspect of the allegory resembles, as well, monadic expression, which is durational and constant. Benjamin's definition of the allegory as speculative, against the profane, distinct, symbol, recalls Leibniz's notion of expression as a 'living mirror of the deity'.[10] God is immanent in every monad, and just so the allegory is characterized by immanence. Similarly, time is immanent in the allegory as history is immanent in Baroque plays. 'History', Benjamin writes, 'passes into the setting'.[11]

The historical aspect of Benjamin's concept of allegory diverges, however, as well, from Leibniz's concept of expression. He describes Baroque allegory as containing a dialectic movement between nature and history, in which history consists of a process of decline rather than progress or salvation. Benjamin describes both allegory and ruin using the term 'natural history', discussed in Chapters 1 and 3.[12] As Adorno and others have shown, Benjamin uses this concept in order to describe nature as historical and history as natural, defining both nature and history through their transience.[13] By defining Baroque allegory as natural-historical, Benjamin highlights the process of slow decline towards finality embodied in the term.

The arbitrary way in which an allegory relates to its meaning in Baroque plays, in which 'any person, any object, any relation can signify any other whatever', is symptomatic of this process of decay of history.[14] Benjamin defines the allegory as a form both empty and overdetermined by meaning, in which the nihilism and melancholy of the Baroque as an age of political and theological crisis is reflected. In Benjamin's later discussions of Baudelaire, whom he views as a modern allegorist, this lack of meaning takes on a different form.[15] Instead of ruins, Baudelaire's allegorical practice brings together commodities, objects whose meaning is determined by their market value. Benjamin describes Baudelaire's poetic gaze as alienated, reflecting the modern fissure between nature and history that had replaced the dialectic movement between the two which had been still possible during the Baroque.[16] If history was immanent to Baroque allegory, Baudelaire's allegories produce a spatial immanence, by internally containing the urban space of Paris.

The reception of Benjamin's theory of allegory, especially during the 1980s when there was a spurt of interest in this concept, has often focused on allegory as a precursor of montage techniques and theories, a procedure of appropriation and composition using disparate elements out of which meaning emerges.[17] Benjamin's personal engagements with prominent members of avant-garde movements during the 1920s and 1930s, the same years he composed his Baroque study, are foregrounded in these accounts. Peter Bürger writes, for example, that 'there is thus nothing forced in the attempt to read Benjamin's concept of allegory as a theory of an avant-gardist (non-organic) work of art'.[18] While I too argue for a link between the procedures of allegory and montage in Benjamin's work, there are also some differences. As mentioned in the previous chapter, Benjamin discusses two forms of montage in different texts, montage as principle and montage as procedure. The allegory is closer to montage as principle, and both bear a stronger connection to Leibniz's philosophy than montage as procedure. As discussed, for Benjamin, the procedure of montage includes a severing and decontextualization of elements. For Deleuze, allegory is rather all about appertaining, belonging to something.

ii Allegory of an allegory: Aristotle's telescope

In Deleuze's brief reading of Benjamin's Baroque allegory theory, undertaken in the last chapter of *The Fold*, he emphasizes the allegory as a 'power [*puissance*] of figuration' rather than a representational technique. The allegory forms part of the schizophrenic excess of folds, the 'all over' fold, which Deleuze describes as 'discovered again in modern art'.[19] This account corresponds to Benjamin's description of the over-determination of meaning in Baroque allegory. Deleuze writes:

> Walter Benjamin made a decisive step forward in our understanding of the Baroque when he showed that allegory was not a failed symbol, or an abstract personification, but a power of figuration entirely different from that of the symbol: the latter combines the eternal and the momentary, nearly at the centre of the world, but the allegory uncovers nature and history according to the

order of time. It turns nature into history and transforms history into nature in a world that no longer has a centre.[20]

Deleuze focuses on Benjamin's distinction between allegory and symbol, emphasizing the temporal aspect of the allegory and the way it mirrors the de-centred, a-harmonic world of the Baroque. He describes the nature-history dialectics within the Baroque allegory, in which nature and history are transformed into one another. Benjamin's allegory is interpreted as an *objectile*: a mannerist, excessive object that overflows its frame like the folds of material that spill outside the frame in Baroque paintings.[21] Deleuze offers an alternative definition of allegory and symbol through the ways in which they relate between concept and object:

> [I]f we consider the logical relation of a concept to its object, we discover that the linkage can be surpassed in a symbolic and an allegorical way. Sometimes we isolate, purify or concentrate the object ... sometimes, on the contrary, the object is broadened according to a whole network of natural relations. The object itself overflows its frame in order to enter into a cycle or a series, and now the concept is what is found increasingly compressed, interiorised, enveloped in an instance that can, if need be, be called 'personal'.[22]

For Deleuze, the allegory describes a relation of envelopment of the concept by the object, or in other words, an immanent inclusion of the concept within the object. In the symbol, on the contrary, the concept is elevated, transcending its object. Appertaining – belonging to a network, cycle or series – is key to allegory, in opposition to the isolation of the symbol.

The allegory is repeatable, then, as part of a series. It functions as an event in that it recalls its previous and succeeding events. It is relational, in that its meaning is constructed through its relations to other allegories in the series, other events. It is also often constructed, in Deleuze's eyes, through the relation between text and image, and the elements that make up an allegory are summed up as 'images or figurations, inscriptions or maxims, and personal signatures or proper names of owners'.[23] Deleuze delineates a paradigmatic Baroque allegory: the world as cone or cupola. Instead of the classical world allegorized as a circle in which everything is drawn towards the centre, Deleuze contends that the Baroque world tends towards the summit or apex of the cone.[24]

Allegory forms part of Leibniz's perspectivist world that has lost its centre. Within this world, Deleuze suggests, unity no longer comes from a centre, but rather emerges from a summit, that itself functions as a point of view.[25] Deleuze's emphasis on the cone is inspired by Leibniz's conic sections, as well as by *Conceptismo*, a Spanish Baroque literary movement known for its use of metaphors and wordplay. Deleuze discusses Emmanel Tesauro's *Aristotle's Telescope [Il cannocchiale aristotelico]* (1654), a key work on *Conceptismo* style, describing the illustration on its Frontispiece as an 'allegory of allegory'.[26] In it, a figure who represents poetry gazes at the sun through a telescope, symbol of the new optical instruments of the Baroque and an allusion to Galileo's studies on sunspots.[27] The telescope is held by a bearded man who is partially hidden in the shadow, identified as Aristotle. A third figure, who represents painting, is seated facing an easel, in the midst of a painting of a cone that hangs suspended over the horizon. The inscription '*Omnis in unum*' – all in one, or multiplicity in unity – is drawn over a conical mirror as an anamorphisms, the conic shape contributing to the legibility of the words.

The Frontispiece allegorizes the role of poetry and painting and their relation to one another and to the science of optics. It functions as an 'allegory of allegory' in illustrating the relational production of meaning through the interaction between the visual and the verbal, as well as in its representation of multiple levels of allegoresis: a painting within a painting depicting a cone representing the Baroque world. Aided by optics, the words, symbolizing the unity of God and the world, attain legibility. 'Unity in multiplicity' is, as we recall, also one of Leibniz's definitions of the monad. The illustration also opposes Classicism and Baroque/Mannerism, by implying that the inscription can only appear with the assistance of Baroque science and art. At the same time modernity is supported by antiquity, as suggested by Aristotle deftly holding the telescope.

The significance of the allegory for Deleuze is in its embodying the form of a concept that is enveloped by the individual.[28] Rather than subject-object relations the allegory presupposes concept-object relations that are philosophically expressed in Leibniz's concept of the individual and the production of individuality through perspectivism. The allegory thus contains, for Deleuze, the seeds of modernity that will transform the monadology into a 'nomadology'.[29] While Deleuze highlights the concept of individuality in relation to modernity, we will see that for Benjamin, modernity is exemplified in the collective body.

iii Image politics

The dialectics between the visual and the verbal, as manifest in the Baroque allegories and emblems analysed in *Origin of the German Trauerspiel*, are also present as a theme in Benjamin's essays that discuss contemporary art and literature. The tensions between exceptionality and repetition, discussed in relation to Benjamin's philosophy of history and Leibniz's concepts of eternal return in Chapter 3, are presented here as well. Before exploring Benjamin's 'dialectical image' and Deleuze's 'crystal-image', we will explore the relationship between image and politics in Benjamin's Surrealism essay, an important transitional text between the earlier *Origin of the German Trauerspiel* and the *Arcades Project*. Although Leibniz is not invoked in this essay, it exemplifies Benjamin's exploration of the unconscious and of liminal, embodied and de-centred perception, which, as we have theorized in the previous chapter, relates to his reading of Leibniz.

In 'Surrealism, the Last Snapshot [*Momentaufnahme*] of the European intelligentsia', published in 1928, Benjamin observes the politicization of the French Surrealist movement between 1919, when it issued its first manifestoes, and its present form – a process he describes as 'pushing the 'poetic life' to the utmost limits of possibility'.[30] This transformation of the movement, accompanied by a shift from 'metaphysical materialism' to 'anthropological materialism', is exemplified, in Benjamin's account, by the role that images play in the Surrealists' work. Benjamin argues that a 'dialectical kernel' of revolutionary potential existed already in the early days of the movement. The radical conception of freedom developed and practised by early Surrealists paved the way for the later politicization of the movement. Thus, in its early stages,

> [l]ife seemed worth living only where the threshold between waking and sleeping was worn away in everyone as by the steps of multitudinous images flooding back and forth; language seemed itself only when sound and image, image and sound interpenetrated with automatic precision and such felicity that no chink was left for the penny in the slot called 'meaning'.[31]

The description of mechanized interchanging of image and sound might be a reference to early surrealist cinematic experiments, inspired by experiences of intoxication through the use of drugs such as hashish, which eroded the boundaries between waking and sleeping, consciousness and unconsciousness. These experiences of liminal states of consciousness, in which boundaries are dissolved, precipitate an upsurge of images and the interpenetration of sound and image. Benjamin may have had in mind films such as Man Ray's *Le Retour à la raison* (1923), composed of photograms and the play of abstract shapes, light and shadow, alongside strips of light on a female torso; or 'Ghosts before breakfast' [*Vormittagspuk*] (1927–8) by Hans Richter. In the latter, time is a protagonist, represented by the face of a ticking clock, its hands moving with uncanny speed, and by fast-forward and reverse motion, which cause the leaves on a branch to sprout in seconds.

Such states also brought about a dissolving of the strictly determined 'individuality' of the artist. The loosening of the 'self' or individuality instigated by stimulants, Benjamin argues, was precisely what allowed them to go beyond intoxication and reach what he names a 'profane illumination'. In other words, the seeds of the revolutionary ideology that will take hold of the movement at its later stages lie in replacing a well-defined concept of 'self' with a 'flood of images'. Rather than merely leading to states of drugged ecstasy, the Surrealists' engagement with dream-like states served to reach beyond religious illumination and attain a 'materialistic, anthropological inspiration'.[32]

The image has a unique role in the interplay of relations that Benjamin delineates between literature, specifically poetry, on the one hand, and political action or activity [*Aktivität*], on the other. Informed by a manifold of avant-garde movements, among them the French Surrealists, Benjamin had already explored the relation between action, writing and image in *One-Way Street* (written in parallel with *Origin of the German Trauerspiel*, published 1928), for example, in the *Denkbild* entitled 'Filling Station':[33]

> True literary activity [*wahre literarische Aktivität*] cannot aspire to take place within literary frameworks; this is rather the habitual expression of its sterility. Significant literary effectiveness [*bedeutende literarische Wirksamkeit*] can come into being only by strict alteration between action and writing; it must nurture the forms that fit its influence in active communities better than does the pretentious, universal gesture of the book – in leaflets, brochures, articles, and placards.[34]

Literature gains its effectivity and force through its interconnection with action. Benjamin draws here on Russian constructivists such as Sergei Tretyakov, one of the leading figures in the Factography movement active in the late 1920s, and originator of the notion of the artist as 'operator', who transforms rather than reflects reality.[35] In order to conserve its powers in politically active communities, it must adopt forms that bring text closer to image, or render it graphically *bildlich*, image-like, such as leaflets and placards.

The image-quality of text renders it political, as stated more explicitly in the *Denkbild* 'Attested Auditor of Books', discussed in Chapter 3.[36] There, as in the essay on Surrealism, both literary and political activity are likened to sources of energy and charged with the dynamics of machinery. In *One-Way Street*, writing becomes politically potent by the edging of script closer to the image, as in Mallarmé's 'Un coup de dés'. This is comparable to Benjamin's discussion of rebus, riddles composed of illustrated letters, in *Origin of the*

German Trauerspiel, in which case the illustrative qualities of the letters hide a second, speculative meaning.[37]

It is the image, photographic or literary, that enables the shift from activity [*Aktivität*] to actuality [*Aktualität*], rendering Surrealistic art truly political. Benjamin describes this shift by referring to Louis Aragon's 'Treatise on Style' (1927).[38] Written during the year in which Aragon joined the French Communist Party, this work is considered an important point in his transition to 'Socialist Realism'.[39] The discussion of poetic style is exemplified in the distinction Aragon draws between metaphor and image.[40] Aragon describes the fleeting quality of the image by its encapsulation of humour, which belongs to a specific time only and is rendered obsolete once this period has passed. When repeated too many times, the joke is ruined. Aragon, and Benjamin following him, alludes to the transgressive function of humour, its ability to cause a stir by enabling directness, saying things that are otherwise unacceptable. The image, charged with the dynamic force of time, must produce shock and upheaval, crumbling the old and ushering in the new. The metaphor, by contrast, is associated by Aragon with the habitual, used and abused that has become a cliché.

Benjamin adopts this distinction in order to differentiate between Surrealist poets and 'poets of the social democratic associations', who write poetry that is 'bursting with metaphors'.[41] Benjamin uses metaphor [*Metapher*] and comparison [*Vergleisch*] interchangeably, making manifest his critique of the metaphor, which functions as a simple comparison. Unlike banal, repeatable metaphors, the Surrealists produce unique images in their art by drawing on dream-images and mental images produced by altered states of being.[42] In 'Dream Kitsch' (1927), Benjamin makes a similar distinction between banal, repeated images, which he designates as 'outer images of things' [*Außenbild den Dinge*] and 'the most objective image of our feelings'. The first are rendered valueless by technological mass reproduction. Worn down by habitual representation, even the appearances of objects in dreams are rendered kitschy, like overused maxims that lose their meaning.

Deleuze, like Benjamin, critiques the metaphor, which he sees as a mode of 'designation' and opposes this to the procedure of 'expression'. But he also takes this idea a step further:

> Kafka deliberately kills all metaphor, all symbolism, all signification, no less than all designation. Metamorphosis is the contrary of metaphor. There is no longer any proper sense or figurative sense, but only a distribution of states that is part of the range of the word. The thing and other things are no longer anything but

> intensities overrun by deterritorialized sound or words that are following their line of escape.[43]

As does Benjamin, Deleuze views the use of metaphor as designating a comparative, symbolic relation. Yet while Benjamin contrasts between the metaphor and the intensive image, charged with the tensions of humour, Deleuze opposes metaphoric figuration to intensity itself.[44] It is the intensive use of language, through the construction of 'machines of expression', which enables metamorphosis and becoming in Kafka's writing. These replace the metaphor, which Kafka 'deliberately kills'.[45] For both Deleuze and Benjamin, then, these intensive modes of expression, which replace the metaphor, are politically effective. For Deleuze, this is a politics of the 'deterritorialisation' of language, words and sounds, within what he terms 'minor literature'.

iv Image and body space

Benjamin targets the moral metaphor that loses its meaning via over-repetition as part of his critique of the failure of the intelligentsia to connect with the proletarian masses. He implies that artists who truly wish to connect with the masses should interrupt their careers and become directly involved in political action, as Tretyakov did, and Benjamin himself did not.

'Image space' is the space of interruption, of shock or emotional reaction caused by impolite and direct words. It is the space in which mental image confronts external action:

> For in the joke too, in invective, in misunderstanding, in all cases where an action [*ein Handeln*] puts forth, and is, its own image, absorbing and consuming it, where nearness views itself with its own eyes, the long sought image space is opened, the world of general and integral actuality [*allseitiger und integraler Aktualität*], where the 'best room' is missing – the space, in a word, in which political materialism and physical creatureliness share the inner man, the psyche, the individual, or whatever else we wish to throw them, with dialectical justice, so that no limb [*Glied*] remains untorn. Nevertheless – indeed precisely after such dialectical annihilation – this will still be an image space and more precisely, a body space.[46]

Action produces image, and itself becomes image, in problematic, affective zones, such as those of the joke, which can be funny and/or offensive, and those of insult or misunderstanding; places in which there is confusion or non-agreement regarding sense and meaning. In these instances, when action discovers its image and takes its form, while also destroying it, 'image space is opened'. 'Image space' demands, then, the uncomfortable in-between zones in which continuity between exterior and interior is created. In image space, the individual is replaced by the collective. Benjamin describes the collective as a body whose members are torn apart. By harnessing the energies of intoxication to the revolution, the Surrealists contributed to the process of collectivization and liquidation of the self, which constitutes a necessary prelude to revolutionary action.

> The collective is a body, too. And the *physis* that is being organised for it in technology can, through all its political and factual reality, be produced only in that image space to which profane illumination initiates us. Only when in technology body and image space so interpenetrate that all revolutionary tension becomes bodily collective innervation, and all the bodily innervation of the collective become revolutionary discharge, has reality transcended itself to the extent demanded by the *Communist Manifesto*. For the moment, only the Surrealists have understood its present commands. They exchange, to a man, the play of human features for the face of an alarm clock that in each minute strikes for sixty seconds.[47]

In the opening of the essay Benjamin depicted the French movement as the current of a river, the energies of which the German critic can observe from his vantage point in the valley.[48] Here, this force becomes that of bodily collective innervation that results in 'revolutionary discharge'.[49] The 'psychic energy' Benjamin referred to in the fragment 'Outline of the Psychophysical Problem', discussed in Chapter 4, is here transformed into the 'collective innervation' of the society as body. The intermingling of the collective as body with its actual, intensive image through the intervention of technology creates a continuity between art and action that may bring about revolution. This has been achieved by the Surrealists' innovative use of technology in their art, reconceiving the body/collective in objects and vice versa.

v Monstrous perspectivism

In the preface to the English edition of *Difference and Repetition*, Deleuze writes:

> A new image of thought – or rather a liberation of thought from those images that imprison it: this is what I had already sought to discover in Proust. Here, however, in *Difference and Repetition*, this search is autonomous and it becomes the condition for the discovery of these two concepts.[50]

We have encountered, in Chapter 2, Deleuze's concept of the dogmatic 'image of thought'.

Above too, the image is conceived of as an incarcerator of thought; the possibility of a 'new', alternative image of thought is raised only to be dismissed. Since Deleuze's project consists of challenging what he calls 'philosophy of representation', the image, in its representational function, is targeted as part of his attack. Yet, as will be shown, Deleuze does not completely do away with visual categories. In fact, he describes a certain kind of perspectivism as an alternative to philosophy of representation. This concept of perspectivism draws on Leibniz's perspectivism yet reformulates it as de-centred and distorted.

Similarly, he will try to redeem a certain type of image that does not confine thought, namely the simulacrum: the distorted image that is not based on semblance. Despite being emptied of its representational functions, the simulacrum is still deemed a kind of image. If we may say that Deleuze puts forward an 'image theory' in this work at all, it is an anti-iconological one; a perversion of the image.

Deleuze's use of the conception of 'simulacrum' forms part of his project, carried out in Nietzsche's footsteps, of 'reversing Platonism'.[51] In *Logic of Sense* Deleuze explains that this project goes beyond the challenging of Plato's distinction between essence and appearance, a challenge already put forth by both Hegel and Kant, and past the distinction between the thing itself and its image, original and copy.[52] Deleuze argues that his reversal of Platonism consists in challenging Plato's oppositional thought, and specifically, in the case of the *Sophist*, the opposition Plato makes between two types of images [*eidolon*]: the copy [*eikon*] and the simulacrum [*phantasma*].[53]

In Deleuze's reading, both types of image are 'claimants' [*prétendants*] that pretend to be what they are not. While the first are 'well-grounded pretenders' that resemble what they intend to copy, the second are 'false claimants, constructed by dissimilitude, implicating perversion'.[54] In other words, the first may be described as true images, while the second as false ones. Deleuze argues that the manifest distinction between *image* and *idea* is only there to ensure the latent distinction between the two types of images: *eikon* and *simulacra*. Like the image, Plato's idea comprises a relation of semblance: something is judged to be 'just' when it resembles the idea of justice. The copy resembles something when it resembles its 'idea', which comprises the relations constituting its essence. It is thus based on intrinsic relations to a model, rather than extrinsic relations to an object.[55] Moreover, the idea is only created in order to track down false images or simulacra, the only type of image that is independent of its referent.

Far from being a degraded copy, Deleuze's conception of the simulacrum differs from the copy in nature: it is an image lacking resemblance.[56] The simulacrum transcends the very distinction between original and copy. Simulacra still produce an illusionary effect of semblance, writes Deleuze; they are demonic, hallucinatory, constructed upon difference. As in the case of an optical illusion, the effect of semblance is produced by including the point of view of the observer, which comprises part of the simulacrum itself.[57]

Deleuze may have attempted to liberate 'thought from those images that imprison it', yet the 'image' is omnipresent in *Difference and Repetition*, as in Deleuze's later works. Deleuze refers to the image of thought in the singular, as he explains, because he does not target specific presuppositions belonging to certain philosophical traditions, but rather, a 'single image in general that constitutes the subjective presupposition of philosophy as a whole'.[58] Naming the presuppositions of philosophy as a single 'image of thought' enables Deleuze to argue that there are presuppositions that are common to all of philosophy and that are, in a sense, visible and apparent like an image. 'Image' is used by Deleuze in order to denote that which is not philosophy, whether it is 'pre-philosophical' or 'non-philosophical'.

Deleuze's attack on the image is more precisely an attack on a specific type of image. His attempt to reverse Platonism thus replicates the uneasy coupling – starting from Plato and dominating the European philosophical tradition – of the image's branding as non-philosophical on the one hand, and the subordination of conceptual thought to the visual field on the other.

Perhaps for this reason the function of the 'image of thought' is reversed in *What Is Philosophy?*, in which Deleuze and Guattari identify it with the 'plane of immanence':

> The plane of immanence is not a concept that is or can be thought but rather the image of thought, the image thought gives itself of what it means to think, to make use of thought, to find one's bearings in thought … The image of thought retains only what thought can claim by right. Thought demands 'only' movement that can be carried to infinity. What thought claims by right, what it selects, is infinite movement or the movement of the infinite. It is this that constitutes the image of thought.[59]

The image of thought cannot be thought, writes Deleuze; once again pairing image with thought only to then distinguish between the two. Like the dogmatic 'image of thought' outlined in *Difference and Repetition*, the 'image of thought' in *What Is Philosophy?* is 'non-philosophical' in the sense that it is non-conceptual. Yet in this case it is not described as a 'dogmatic', 'pre-philosophical' image. Rather, the 'Image of thought' maintains a crucial function in relation to thought and the construction of concepts. It constitutes a 'living mirror' of thought; a movement in which thought is folded, unfolded, and refolded to infinity. It is an 'open whole', a 'One-All' [*Un-Tout*] that comprehends all concepts on a 'plane' or 'horizon' from which they can be selected.

The 'plane of immanence' harbours a flux of concepts constantly developing, mutating through shifts in their degree of intensity. This understanding of philosophy as a conceptual 'life' originates, as Deleuze and Guattari concede, in Leibniz and Bergson:

> The grandiose Leibnizian or Bergsonian perspective that every philosophy depends upon an intuition that its concepts constantly develop through slight differences of intensity is justified if intuition is thought of as the envelopment of infinite movements of thought that constantly pass through a plane of immanence.[60]

As mentioned, Deleuze designates what is unique about the simulacrum by using the category of the 'modern'.[61] The term 'modern' – which Deleuze uses, in *Difference and Repetition,* in relation to art, literature, 'thought' and mathematics, and later in relation to cinema – is described as that which has the force to bring about the collapse or failure of representation. Moreover, it is modern art that leads the way in challenging representation, and philosophy must follow its lead. The challenge modern art poses to 'representation' is by way of what Deleuze describes, once again using visual terminology, as a kind of perspectivism:

> Infinite representation may well multiply points of view and organise these in series, these series are no less subject to the condition of converging around the same subject, upon the same world … By contrast, when the modern work of art develops its permutating series and its circular structures, it indicates to philosophy a path leading to the abandonment of representation. It is not enough to multiply perspectives in order to establish perspectivism. To every perspective or point of view there must correspond a work with its own self-sufficient sense: what matters is the divergence of the series, the decentring of circles, the 'monster'.[62]

Deleuze draws on Leibniz's perspectivism, yet at the same time criticizes Leibniz's conception of infinite representation for the fact that all points of view represent one and the same world. Deleuze suggests that in modern literature an extreme version of perspectivism occurs, in which the multiple perspectives do not converge into one world or one work: 'it is as if a completely distinct landscape corresponded to every point of view.'[63]

In this perspectivism, which becomes a monstrous 'chaos', series diverge and narrative circles are de-centred. The simulacra produced by modern literary works such as Joyce's *Finnegans Wake* and Mallarmé's *Livre* are described, then, as a kind of chaotic perspectivism, in a manner that anticipates Deleuze and Guattari's description of art as 'a composition of chaos' in *What Is Philosophy?*[64]. By mentioning Mallarmé, Deleuze gestures at the figure of the 'dice-throw' and questions of chance, repetition and predetermination discussed in Chapter 3.

The concept of chaos, which draws on modern chaos theory but is also defined as 'a *virtual*, containing all possible particles and drawing out all possible forms', draws a line from Leibniz and Early Modern science to twentieth-century science, from the 'dice-throw' to Heisenberg's principle of uncertainty.[65] It contains a critique of the assumptions of truth and knowledge that underlie the development of probability theories by early modern thinkers. While Leibniz believed that all series converge in God, who holds perfect knowledge and to some extent determines the future through the principle of pre-established harmony, Deleuze's monstrous perspectivism is based on chaos and therefore assumes elements of uncertainty and unpredictability.

2 Between the dialectical image and the crystal-image

i Crystal-image and doubled time

Deleuze and Benjamin's most renowned image-concepts, the 'crystal' and 'dialectical' image, are at the same time counter-image concepts, through which both authors critique the representational hegemony, and as such they enfold and enact 'mannerist perspectivism' and 'historical perspectivism' respectively, as will be explored in the following.

The *Cinema* books contain a 'taxonomy of images', as Deleuze writes in the preface to the French edition of *Cinema 1*.[66] Deleuze uses the plural form to describe a range of 'sign-images' that are different from the two-image conceptions discussed hitherto, the dogmatic 'image of thought' and the 'simulacrum'. Moreover, unlike the 'image of thought', described by Deleuze as pre-philosophical, the 'images' he discusses in the *Cinema* books function as philosophical concepts.

The key inspirations for Deleuze's conceptualization of the image and its relation to time in these works, he writes, are the philosophy of Bergson and the semiotic theory of Charles Sanders Peirce. Yet Leibniz's philosophy, I maintain, plays a larger role in these works, especially in *Cinema 2*, than betrayed by the occasional direct references to his philosophy.[67] Already in *Cinema 1* there are intimations of the manner in which cinema enacts a kind of perspectivism, which become more explicit in *Cinema 2*. Deleuze describes, for example, in the opening pages of *Cinema 1*, the evolution of cinema as relying on the 'emancipation of the viewpoint' [*prise de vue*, which is the term for 'shot' or 'shooting'] once the apparatus of shooting [*appareil de prise de vue*] was separated from that of projection.[68] This transformed the shot from a spatial into a temporal category, he writes. By using the term *prise de vue*, which is semantically close to *point de vue* (point of view), Deleuze refers to the emancipation of the shot from the apparatus that produced it and also to the separation between the point of view of the cinematographer, and that of the viewer. The term 'projection', moreover, links his discussion of cinema to that of projective geometry.

The two titular concepts of the *Cinema* books, 'movement-image' and 'time-image', are directly appropriated from Bergson's *Matter and Memory*.[69] Although Bergson wrote little on cinema and when he did, it was in rather unflattering terms, Deleuze found insights for his understanding of cinema in Bergson's philosophy of time and the particular manner in which he understood perception, which was informed by Leibniz, and which further explains the latent presence of the latter in the *Cinema* books. Bergson conceived of the image as an elementary particle of the universe, which consists of an aggregate of images.[70] Unlike Leibniz's monads that are closed off from one another, Bergson's images constantly act and react upon one another. As in Leibniz, psychic and physical realities are intermingled. Bergson describes the relation of perception and reality as a continuous and analogous relation of part and whole, in a manner not unlike the relation of expression Leibniz designates between the perceiving soul and the universe. Moreover, Bergson and subsequently Deleuze rely on Leibniz's theory of the unconscious for their understanding of consciousness as a secondary process which follows a pre-conceptual moment of being affected in the world. The primary stage of sensation, Bergson maintained, is followed by a secondary one of consciousness, to which processes such as recognition and perception belong.

Deleuze defines, in *Cinema 1*, the 'movement-image' accordingly as the 'a-centred set [*ensemble*] of variable elements which act and react on each other'.[71] While the 'movement-image' represents time in an indirect manner, deriving time from movement, time-image, elaborated upon in *Cinema 2*, represents time directly, transcending the movement-image. It consists of a 'new image' of post-war cinematics, inaugurated by neo-realist cinema. It is a 'readable' [*lisible*] and thinking image [*image pensant*], yet also one which 'commands *false movement*'.[72]

The 'crystal-image' [*L'image-cristal*] makes its first appearance in the fourth chapter of *Cinema 2*, titled 'The Crystals of Time'. Deleuze delineates it as 'the coalescence of the actual image and *its* virtual image, the indiscernibility of two distinct images'.[73] The 'crystal-image' constitutes a 'point of indiscernibility' in which actual and virtual are distinct and yet, paradoxically, cannot be distinguished one from another. It combines the 'movement-image', which Deleuze describes as actual, and the 'time-image', which he describes as virtual.[74] It is informed, as Deleuze concedes, by Bergson's definition, in *Matter and Memory*, of 'circuits' created between the purely optic and the sonorous, that is, between the actual image and a kind of 'memory image'.[75] While informed by Bergson, as well as Jean Ricardou's theory of the *Nouveau Roman*, the term 'crystal-image' itself is not Bergson's.

'The crystal is expression. The expression moves from the mirror to the seed',[76] writes Deleuze, defining the crystal as enacting a double movement of actualization and mirroring. Recall that the two key metaphors of expression Deleuze describes in *Expressionism in Philosophy*, as discussed in Chapter 1, were the mirror and the seed. In the first case the expressed is reflected as a mirror-image of the expresser, as in Leibniz's 'living mirror' metaphor. Deleuze quotes Ricardou, who describes the crystal as a 'mobile mirror that unceasingly reflects perception in memory'.[77] He gives as an example the famous ending scene from *The Lady from Shanghai* directed by Welles, which takes place in a theme park maze of mirrors involving the main characters Elsa, her husband, Arthur, and her lover, Michael. 'A perfect crystal-image where the multiple mirrors have assumed the actuality of the two characters who will only be able to win it back by smashing them all, finding themselves side by side and each killing the other.'[78] The mirrors physically reflect the characters, fragmenting and replicating them, and at the same time assume their actuality, expressing the fate of Elsa and Arthur, who will die when the mirrors are shattered.

In the second case, the expressed is involved or implied in the seed. In cinema, the mobile mirroring is affected by instances of *Mise en abyme* such as a theatrical play, an image or a film within a film. In this case, the reflection is imperfect, the crystal is more opaque than transparent. Expression by way of the seed is carried out by films that take as their object the process of film production, especially when this process is unsmooth and difficult, as in Fellini's *8½* (1963), in which the director of the film struggles with creative blocks, or Wim Wenders' *The State of Things*, in which the filmmaking process is aborted.[79]

> The crystal-image has these two aspects: internal limit of all the relative circuits, but also outer-most, variable and reshapable envelope, at the edges of the world, beyond even moments of world. The little crystalline seed and the vast crystallizable universe: everything is included in the capacity for expansion of the collection constituted by the seed and the universe. Memories, dreams, even worlds are only apparent relative circuits which depend on the variations of this Whole.[80]

Crystallization equals expression: the virtual image, composed of memories and dreams, expresses the whole, the universe and its variations, the latter being contingent upon its expression within the 'seed'. The crystal-image amalgamates Bergson's understanding of time, according to which past and present, virtual and actual images co-exist, connected by 'circuits' – with Leibniz's conception of expression. As noted, Bergson's understanding of the relation between mind and matter was informed by Leibniz. He directly refers, in *Matter and Memory*, to Leibniz's notion of monads as centres of force and living mirrors of the universe.[81] In a sense, Bergson's notion of circuits takes Leibniz's expression a step further; in lieu of continuity between the expresser and the expressed, they constitute a full circle, like a Mobius strip. As Deleuze writes,

> What constitutes the crystal-image is the most fundamental operation of time: since the past is constituted not after the present that it was but at the same time, time has to double itself [*se dédouble*] in two at each moment as present and past, which differ from each other in nature, or, what amounts to the same thing, it has to double the present in two heterogeneous directions, one of which is launched towards the future while the other falls into the past … Time consists of this split [*scission*], and it is this, it is time, that we *see in the crystal*.[82]

The crystal-image appears not in space but in time. It is, moreover, constituted through the 'fundamental operation of time', which Deleuze understands, following Bergson, as a doubling of the present that takes place during each and every instant, each of the 'doubles' oriented in a different direction: the first towards the future, and the second towards the past. The split between the two constitutes time, and it is this split that may be seen within the crystal. The co-existence of the double present – past-present and present-present – is appropriated, as Deleuze admits, from Bergson's *Mind-Energy*.[83]

While Bergson describes the co-existences of past and present, in Leibniz, space consists of the order of co-existence while time is that of succession. For Deleuze, the division or doubling of the present at every instant into two 'presents' – one facing the past and one approaching the future – is a critical moment in Bergson's thought, embodied in the conception of the 'crystal-image'.[84]

ii Dialectical image and now time

Benjamin's concept of the 'dialectical image' is formulated, like Deleuze's 'crystal-image', at a late stage of his work, although as we have seen, Benjamin's other image-concepts anticipate the dialectical image. The context of its appearance in the *Arcades Project* is quite different from that of Deleuze's crystal-image, which is discussed specifically in a cinematic context. However, Benjamin and Deleuze's image-concepts operate in a similar manner in relation to time, and similarly emerge at the conjunction of visible and verbal registers: dialectical images are to be encountered 'in language', writes Benjamin, while Deleuze's image-concepts form part of a 'semiology of the cinema'.[85] While some have argued that Benjamin's dialectical image is primarily linguistic, I do not believe this to be the case; rather, the intermingling of the visual and linguistic dimensions within the image are unique to both the dialectical and the crystal-image, bringing them closer together.[86] This conjunction of visual and verbal, image and word are evidence of the shared origins of these concepts in the Baroque allegory and mannerist style.

We have seen that temporality is a significant aspect of the allegory, both in Benjamin and Deleuze's interpretations. In Benjamin's dialectical image, as in the crystal-image, there is a doubling of time. Here the doubled times are not past and present but rather 'what has been' and 'now'. To reprise Benjamin's memorable depiction of the dialectical image,

> It's not that what has been casts its light on the present, or what is present its light on the past; rather image is that wherein what has been comes together in a flash with the now to form a constellation. In other words, image is dialectics at a standstill. For while the relation of the present to the past is a purely temporal, continuous one, the relation of what has been to the now is dialectical: it is not progression but image, suddenly emergent.[87]

Both crystal and dialectical images contain a critique of chronological, progressive time through its doubling. In the dialectical, as in the crystal-image, there is a coalescence of actual and virtual; the actual 'now' constellates with the virtual memory of 'what has been'; these remain distinct and yet at the same time they are mutually indiscernible.

There is a second doubling of time in Benjamin, in which historical time becomes distinct from progressive time. For Benjamin, for historical time to come into being, continuous time must stop. The image surges, then, outside of continuous time. Its position beyond linear time, moreover, enables the construction of *historical* time. Deleuze similarly differentiates between chronological and non-chronological time, the foundation of the latter which is visible within the crystal-image.[88] Yet, in Deleuze, movement is not arrested for the image to emerge; time 'splits' but does not stand still, nor does it offer an external vantage point vis-à-vis time, as does the dialectical image. Benjamin's construction of history is thus contingent upon an a-temporal point of view from which history as a totality may be observed.[89]

When the dialectical image is read, as noted in previous chapters, the image is in 'the now of its knowability [*Jetzt der Erkennbarkeit*]'.[90] The experience of the 'now' is linked through this concept with the possibility of knowing, a 'knowability', as Benjamin defines it. Benjamin's critique of subject-based conceptions of knowledge, discussed in the first chapter, is manifest in the concept of the 'now of knowability', which defines an objective process of knowing, one that, forming part of the image, is not contingent upon a specific 'reader'. Rather, it is dependent on the arrival of a fleeting 'now' in which it can be recognized: every 'now' has another 'now of its recognisability', and when the two are constellated, an image emerges.

Deleuze has something quite different in mind when he describes the 'time-image' as being 'read'. Reading an image means, for Deleuze, paying attention to its sonorous elements, not just the visible data it contains. He describes a process in which the image became readable with the shift from silent to sound films. The addition of speech to the cinematic image contributed to an equivocity of meaning that demands a process of reading in the sense of deciphering a text. The intricacies of relations and meanings within Antonioni's *Story of a Love Affair* (1950), for example, according to Deleuze, may only be deciphered through a process analogous to that of reading.[91] Deleuze's images-signs are readable in the sense that they can be decoded as a sign system, while for Benjamin the image may be 'known' or 'recognised'.

Benjamin departs further from Bergson, and hence from Deleuze, in his definition of the dialectical image as constructive of *historical* time, which he defines as the 'time of truth'.[92] The notions of 'time' and 'history' are opposed to one another in Benjamin's later writing; the first, which he also refers to as 'chronological' or 'progressive' time, denotes a causal chain of events in which all moments share the same significance. He describes this first time as homogenous and empty. 'Historical time', on the other hand, is described as constructed through a kind of montage of dialectical images, or, in other passages, as in itself 'imagistic' [*bildhaft*].[93] That is, while chronological, linear time lacks form, historical time is both formed and informed [*gebildet*].[94]

As elaborated in Chapter 3, Benjamin's concept of the dialectical image is defined as intensive, and as such contains fluctuating forces. Accordingly, in the construction of historical time, as opposed to the sheer amassing of data, 'the different epochs of the past are not all affected in the same degree by the present day of the historian'.[95] That is, different moments of the past carry varying degrees of significance that are contingent upon the historian's perspective. As opposed to 'empty', homogenous time, history or historical time is gradated intensively, 'full' of opposing energies that delineate its form.

In 'On the Concept of History', the last text Benjamin wrote in his lifetime, in 1940, which is composed of eighteen 'theses' on the concept of history, historical time is also termed 'messianic time'.[96] Following Löwy, I view the text as informed by a kind of elective affinity between Marxism and theology, rather than privileging one of these approaches over the other, or pronouncing them as categorically contradictory, as some previous commentators have argued.[97] He contends rather that revolution and redemption are not only inseparable but in fact reinforce one another.

'History is the object [*Gegenstand*] of construction whose site is not homogenous time, but time filled by "now time" [*Jetztzeit*].'[98] The concept of 'now time' denotes the manner in which past epochs are charged intensively with the forces of the present. Rather than merely the 'present', 'now time' designates the time of the dialectical image. Its structure is monadological, like that of the dialectical image and the object of history: 'Now time, as a model of messianic time, comprises the entire history of mankind in a tremendous abbreviation.'[99] The abbreviation enacted through 'now time' includes virtual history past and future. The ratio between 'now time' and the history of mankind corresponds to that between the history of mankind and the history of the universe.[100]

Historical time, in other words, is constructed from minute, seemingly insignificant particles of 'now times', each of which, however, virtually includes the entire history of mankind in abbreviated form. In a version of the text known as the 'Hannah Arendt Manuscript', reproduced from a manuscript Benjamin gave Arendt before his death, the above-quoted passage on the abbreviation of 'now time' is followed by the crossed out sentence 'so that the truth may be accounted for'.[101] Alongside the discussed passage in the *Arcades Project* in which historical time is described as the 'time of truth', these references make manifest the metaphysical aspect of historical time, which at a later date Benjamin partially dismissed.

As has been noted, the opposition Benjamin creates, in 'On the Concept of History', between chronological time, on the one hand, and historical or messianic time, on the other, is prefigured in his dissertation on 'The Concept of Criticism in German Romanticism' (1919). Therein Benjamin designates the essence of Romanticism as messianic, contrasting 'qualitative temporal infinity' [*qualitative zeitliche Unendlichkeit*] with the 'empty infinity of time' [*leeren Unendlichkeit der Zeit*]. The first is 'an infinite process of fulfilment, not a mere becoming', while the second comprises 'progress into the void'.[102] The task of Romantic poetry is viewed as a Leibnizian one, progressing towards fulfilment or perfection within a qualitative infinity, rather than a neo-Kantian progression into empty infinity. It is, according to Benjamin, an infinite yet determinate task.

In the 'Theological-political fragment', probably written in 1921–2, Benjamin similarly associates the messianic with a gradated force, in an enigmatic image of two oppositional arrows, one representing messianic intensity and the other the realm of 'the profane'.[103] These two forces, running in opposite directions, nevertheless empower one another. As Benjamin argues, 'just as a force, by virtue of the path it is moving along, can augment another force on the opposite path, so the secular order – because of its nature as secular – promotes the coming of the messianic kingdom.'[104]

This image of messianic and secular times as split and running in two directions, a double-powered intensity, reminiscent of the doubling of time in Bergson and Deleuze, in a sense anticipates the doubling of time in Benjamin's dialectical image. Therein 'present' and 'past' receive a doubled version as 'what has been' and the 'now'; the dialectical relation between the latter pair echoes the two arrows that each augments the force of other. Moreover, the link between messianic and historical time, explicit in 'On the Concept of History', is already made in the fragment, in which Benjamin defines the relation between the messianic and the secular order as 'one of the essential teachings of the philosophy of history'.[105]

The contrast between qualitative-messianic time and progressive 'empty' time informs 'On the Concept of History' as well. Yet rather than simply being directed towards fulfilment, history is heading towards catastrophe and at the same time contains the possibility of redemption or revolution. Messianic expectation is described as a 'force', albeit one that is 'weak' [*schwache messianische Kraft*].[106] This 'weakness' results from the fact that 'redemption is anything but assured; it is merely a slim possibility, which one has to know how to grasp'.[107] In addition, it is the interruption of this force, the 'messianic arrest of events', which brings about a 'revolutionary chance in the fight for the oppressed past'.[108] Just as messianism is in a sense fulfilled, or actualized, by the coming of the messiah whom, through the fact of his arrival, interrupts messianic anticipation, so the interception of messianic force actualizes revolution bringing about redemption.

The German philosopher Hermann Lotze's *Mikrokosmus* (1856–64) may have served as mediator of Leibniz's ideas and concepts that Benjamin uses in his late writing, years after he had referred directly to Leibniz's texts.[109] In 'On the Concept of History' Benjamin quotes Lotze on happiness as relative to the times in which one lives. 'In other words', Benjamin writes, 'the idea of happiness is indissolubly bound up with the idea of redemption. The same applies to the idea of the past, which is the concern of history. The past carries with it a secret index by which it is referred to redemption'.[110]

In 'Convolut N' of the *Arcades Project*, Benjamin remarks on the link between Lotze's critique of progress and his understanding of redemption, defined as 'the thought of the preservation and the restoration of all things'.[111] The concept of redemption denotes, for Benjamin, Christian apokatastasis as well as Jewish redemption or *Tikkun*. Lotze combined a critique of historical progress with a notion of redemption, or apokatastasis, which, like that of Origen and Leibniz, consisted of a vision of eternal return in which souls are both preserved and restored, denoting, aside from the restitution of souls, an emergence of the new.[112] In 'Convolute N' Benjamin quotes Lotze's description of the form taken by the course of history as 'spirals – some prefer to say epicycloids'.[113] In Benjamin's recurring engagement with conceptions of eternal return, he puts to the test such ideas of the repetition of history against visions of progress, while insisting that repetition includes modification, as discussed in Chapter 3.

In the third 'Thesis', this spiralling form of history is described as its 'citability', the possibility of repetition of historical events through their citation. Redemption is here defined as the salvaging of the details of history from forgetfulness, by committing all historical events, large or small, to remembrance [*Eingedenken*]. In 'Convolut N' Benjamin also describes history as a 'form of remembrance', to which history owes its incompleteness: 'What science has determined, remembrance can modify'.[114] In this case Benjamin uses the notion of history as remembrance in order to argue, along with Horkheimer, for the incompleteness of history. This notion of remembrance as enabling the intervention of the past in the present may very well have been appropriated from Bergson, although he is not mentioned in this passage.[115]

Yet, Benjamin was also critical of Bergson's theory of time as manifest in a brief passage in 'On Some Motifs in Baudelaire' in which he targets Bergson's concept of duration. Benjamin criticizes Bergson for arguing that the summoning of memory is a matter of choice, and as such remains subjective. Applying Proust's term '*mémoire involontaire*', Benjamin designates a temporal experience [*Erfahrung*] in which content from an individual's past combines with material from the collective past, thereby critiquing Bergson's '*mémoire pure*'. In other words, Benjamin faults Bergson's philosophy of time for its subjective and a-historical nature. In this regard, he echoes Horkheimer, who commended Bergson's differentiation between lived and abstract time, yet argued that he remained on 'the threshold' – which he could not cross – 'of dialectics'.[116]

iii Deleuze and Benjamin on montage and montrage

Both the 'dialectical' and the 'crystal'-image enact a montage of sorts: a perspectivized construction that is at once internal to the image, and intervallic, between images. We have discussed how within the dialectical image several points of view are constellated or 'montaged'. Deleuze describes a parallel operation of montage within the 'time-image'. In modern cinema, Deleuze writes, montage has, in a sense, entered the image itself: 'The elements of the image already implicate the montage.' This was only made possible under the conditions of the 'time-image'.

Deleuze describes a process in which, in post-war cinema, the classical 'sensori-motor schema' he described in *Cinema 1* was shattered. What had changed is cinema's self-perception as artifice. While it was always made up of 'aberrant movements and false continuity shots', it took modern cinema to proclaim itself as a virtual image, embodying the phantom that had haunted cinema ever since its inception.[117]

Montage played an important role in this shift. In classical cinema, montage was an operation performed upon movement-images. By contrast, in modern cinema,

> montage has changed its meaning, it takes on a new function: instead of being concerned with movement-images from which it releases [*dégage*] an indirect image of time, it is concerned with the time-image, and releases from it the relations of time on which aberrant movement must now depend. To adopt a word of Lapoujade's, montage has become 'montrage'.[118]

If classical montage extracted an indirect image of time from movement-images, 'mobile sections of duration', as Deleuze defined them, modern montage releases from the time-image – 'duration images, change images, relation images … which are beyond movement itself' – relations of time.[119] In a lecture, he uses slightly different terminology: modern montage, or montrage, is an operation that '*determines* the relations of time within the time-image'.[120] In any case, we are dealing with a process of constructing temporal relations from, and in, 'images' that are temporal in themselves, rather than simply aggregating sections of duration together.

It is not that the procedure of montage itself had changed, or that it is used more or less abundantly, Deleuze specifies. Montage continues, in post-war times, to be the most essential cinematic procedure, yet it is no longer the same. In classical cinema, montage had consisted of a 'vertical construction', linking images successively to each other, while in modern cinema, it functions within the image, for example, by variations in the depth of field.[121] The question guiding montage is no longer how the images connect, but rather, 'what does the image *show*?' [*qu'est ce que montre l'image?*].[122] Montage is no longer distinct from the image; therefore, the time-image could have been termed a 'montage-image'.

Deleuze describes, in *Cinema 2*, two 'regimes' that pertain to pre- and post-war cinema: the 'organic' and the 'crystalline', which he also calls the 'chronic' and 'cinetic' regimes respectively, implying that the crystalline regime is the truly cinematic of the two. The shift from classical to modern cinema echoes a shift in the history of philosophy that Deleuze detects in *Kant's Critical Philosophy*: in ancient philosophy, Deleuze writes, time was subordinate to movement, while today, time is 'out of joint', which signifies 'the reversal of the movement-time relationship. It is now movement which is subordinate to time'. Liberated, time presents itself as an 'empty and pure form'.[123]

Pure time, writes Deleuze, drove the notion of truth to a head. This crisis is made manifest in the 'contingent futures' paradox: if it is *true* that a battle *can* take place tomorrow, how do we avoid one of the two following possibilities: either the impossible proceeds from the possible, since if the battle takes place it is no longer possible that it *can* take place, or the past is not necessarily true, since the battle might not take place. Deleuze refers to Leibniz's solution to the paradox in the *Theodicy*: the battle can take place in one world and not in another, the two worlds are possible, yet are not 'compossible', or in other words, they cannot be created together, since their joint existence would imply contradiction. Leibniz rescues truth through the concepts of 'compossibility' and 'incompossibility', which determine not what is possible but what is *jointly* possible.[124] This means, however for Deleuze, that in Leibniz's universe of multiple worlds, in which narrative no longer claims to be truthful, it becomes essentially falsifying.[125]

Montage within the 'crystalline regime' extracts a falsifying movement out of pure time. Within the direct image of time – an open relational whole – the power of the false is the determinant of relations. 'Narrative becomes temporal *and* falsifying at exactly the same time', writes Deleuze, that is, in order to become temporal it must become falsifying.[126] The 'time-image' and therefore the 'crystal-image', which consists of, in a sense, the ultimate 'time-image', are contingent upon this falsifying movement.

Deleuze refers here to Nietzsche's 'will to power' that substituted truth for the creative force of the false. We have seen that in *Difference and Repetition* Deleuze similarly argues that Leibniz's multiplicity of incompossible worlds should be transformed into eternal returns within this world – returns that are contemporaneous with that which they cause to return.[127] Deleuze opts time and again for a Nietzschean version of Leibniz's doctrine of infinite worlds. This does not mean, however, that Deleuze amalgamates Leibniz's 'mannerist' perspectivism with Nietzsche's perspectivism, the latter which Deleuze describes in *Nietzsche and Philosophy* as a 'total critique', according to which there are no moral facts or phenomena, but only their moral interpretation.[128] Nietzsche is invoked by Deleuze in order to disrupt an aspect in Leibniz's philosophy which he cannot accept, in a manner similar to that in which Deleuze reads Nietzsche against Kant. One such unacceptable aspects of Leibniz's philosophy, for Deleuze, is his understanding of truth, which results from his theological worldview.

As we have seen, Benjamin's concepts of truth in the Foreword to *Origin of the German Trauerspiel*, and in the *Arcades Project*, are forged against classical conceptions of truth as timeless on the one hand, or as based on subjective intention on the other.[129] Benjamin's concept of truth, linked to that of origin, was informed by Goethe's *Urphänomen*, which describes the appearance of totality within an individual.[130] While in the Foreword what is at stake is the truth of the work of art, in the *Arcades Project* the 'point of explosion' within the dialectical image discloses 'true historical time, the time of truth'. The structure of the 'time of truth' enacted by the principle of montage is monadic – it includes a totality of history in every individual moment.

This aspect of Benjamin's notions of montage and the 'dialectical image' appear to be furthest from Deleuze's conceptions of montage enacted in the 'crystalline regime' and by the 'crystal-image'. The crystalline regime, we recall, is concerned with post-war cinema. The films Benjamin viewed belong to what Deleuze terms the 'organic regime'. For Benjamin, film produced by image-montage was the opposite of an 'organic' artwork. As he writes, 'the finished film is the exact antithesis of a work created with a single stroke. It is assembled from a very large number of images and image sequences that offer an array of choice to the editor.'[131] In this case, the editor chooses between the images, mediating the material for the viewer.

Yet for Benjamin, as mentioned, the term 'montage', especially in earlier years, was informed by a variety of media, including Henry James' and Döblin's literary montage, John Heartfield's photomontage and cinematic montage in Soviet cinema from the 1920s.[132] In the *Arcades Project* he famously defines his work as 'literary montage' in which 'I needn't *say* anything. Merely show'.[133] That is, for Benjamin, in literary montage meaning is not *said*, or stated; the author merely selects and exhibits material. The *Arcades Project* itself adheres to his definition of a chronicle in 'The Storyteller', lacking the mediation of commentary, providing the reader with unprecedented freedom. This does not imply that the fragments bear no meaning at all, since as we stated, meaning is constructed relationally. Nor am I suggesting that all fragments share the same significance, but rather that they are not organized according to a specific hierarchy of meaning.[134]

In 'The Author as Producer' (1934) and the Artwork essay, Benjamin's description of montage as a 'procedure' [*Verfahren*] (rather than as a 'principle' as he refers to it elsewhere) seems to refer more directly to cinematic montage. Benjamin's interest in Soviet cinema may have been sparked by his visit to Moscow in 1926–7, where he watched Vertov's *One Sixth of the World*, Eisenstein's *Battleship Potemkin*, Pudovkin's *Mother* and *By the Law* directed by Kuleshov.[135] He was no less interested, however, in *Mickey Mouse* cartoons, writing that 'in these films, mankind prepares to survive civilisation'.[136] These works, as well as constructivist theory, led him to proclaim his 'the age of montaged [*montierbar*] artwork'.[137] There is no evidence that Benjamin had read Eisenstein's theoretical works, discussed by Deleuze in *Cinema 1*. However, Pudovkin's *Filmregie und Filmmanuskript* (1928) is referenced in a note to Benjamin's Artwork essay, the first draft of which was written in 1935.[138] Benjamin quotes Pudovkin, who writes that 'to connect the performance of an actor with an object … is one of the most powerful methods of cinematic construction'.[139] Pudovkin's theory, which defines montage as a tool for guiding and constructing the viewer's 'psychological state', no doubt found its way into Benjamin's conception of montage in the 'Artwork' essay, alongside the perceptual psychologist and film theorist Rudolf Arnheim's Gestalt psychology ideas presented in *Film als Kunst* (1932).

Deleuze's theory of early cinematic montage, unfolded in the third chapter of *Cinema 1*, defines montage as the determination of an open, relational whole of time, or duration. This determination or 'releasing' of the image of time from the ensemble of active and reactive elements that constitute the 'movement-image' is performed 'by means of continuities, cutting, and false continuities [*faux accords*]'.[140] While Deleuze admired Eisenstein's revolutionary cinematic reformulation of dialectics, whereby the latter redefined 'organic' composition as motivated by dialectical forces, he also criticizes Eisenstein's theory of dialectical montage, in which montage is defined as the 'idea' of the film; the film as a whole constituting the 'object' of montage. The whole should rather be primary to its parts, according to Deleuze. Benjamin's understanding of the task of the film-editor, on the other hand, is comparable to that of the chronicler, as described in the 'Storyteller' essay: the selection of and display of materials, out of which meaning emerges. The meaning of the whole, then, depends on the juxtaposition of the parts, much like in Soviet montage theories, yet when brought together, the parts do not constitute an organic whole.

Benjamin's description of montage in the Artwork essay highlights the role of the cinematic apparatus in shaping the perception of the audience. The essential process of film production, that of montage, is determined, he writes, by intervention [*Eingreifen*]. Benjamin gives an example of an actor who is supposed to look startled by a knock at the door. If the director is not satisfied with his performance, he can manipulatively create it by having a shot fired behind the actor and filming his startled expression.[141] The 'test performance' of the actor in front of the apparatus is exhibited to the spectator. The apparatus thus separates the actor's performance, which remains a recorded 'test', from the audience in the cinema.

Although different in many ways, both Deleuze and Benjamin's concepts of montage are marked by Leibniz's cinematic theory of perception, with its artificial devices of perspective (discussed in Chapter 4) and selection among virtual worlds. In Benjamin's case, there are two concepts to be considered: his *principle* of montage bears a closer relationship to the Baroque allegory and Leibniz's expression, while the *procedure* of montage is more closely informed by Soviet films and filmmakers, while in Deleuze we have distinguished between montage and montrage. Deleuze and Benjamin each utilize Leibniz's perspectivism to traverse the path from monad, through allegory, to montage, demonstrating the ways in which Leibniz's theory lends itself to the modern cinematic medium.

3 The crystal pyramid, Leibniz's multiverse

We have noted that Deleuze's main critique of Leibniz, whom he describes in *Difference and Repetition* as a thinker of the identical, targets the 'limitation' of Leibniz's philosophical system, manifest in his contention that the infinity of series that make up the universe converge towards God. Deleuze seeks to surpass these limitations, as we discussed in Chapter 3, through what he views as Nietzsche's Dionysian, open-ended eternal return.

Deleuze suggests that diverging series were only imaginable after Leibniz's time, during the neo-Baroque. In order to avoid diverging series existing side by side, Deleuze writes, Baroque thinkers like Leibniz relegated each to different worlds.[142] 'With the neo-Baroque, with its unfurling of divergent series in the same world, comes the interruption of compossibilities on the same stage, where Sextus will rape and not rape Lucretia, where Cesar crosses and does not cross the Rubicon.'[143] Although this modern 'restaging' of Baroque drama can only happen in the neo-Baroque, Leibniz comes closest to its articulation in his description, in his Theodicy, of the 'crystal pyramid', as Deleuze calls it.

The foundational role of Leibniz's perspectivism for modern cinema and literature is embodied in the myth that seals the *Theodicy*, described by Deleuze as 'a source of all modern literature'.[144] Leibniz's version of the myth is also a perfect example of folded architecture. He uses expression, that is, the folded or nested structure of the monad, as a principle of narrative construction in which 'layers' of dialogue nest in one another. Its moral supports Leibniz's argument, in the *Theodicy*, for the mutual existence of divine justice and human freedom.

In Deleuze's words, a 'crystal pyramid' plays a central role therein.[145] He also describes the pyramid as the reversed Baroque cone, discussed earlier. In fact, the parable unfurled at the closing section of the *Theodicy* describes a pyramid, though not a crystal one. In Leibniz's description of the 'palace of the fates' at the centre of the fable, he writes that the halls within the palace 'rose in a pyramid'; that is, what looked like a palace from the outside became a pyramid from within, in which every hall transformed into a complete universe, even including its own sun and stars.[146] Like the inscription written on the cone on the Frontispiece of Tesauro's *Aristotle's Telescope* discussed earlier, it is an anamorphosis, or optical illusion, that changes its appearance according to one's point of view.

The way in which Leibniz describes the pyramid suggests that it is semi-transparent and semi-permeable.[147] Yet it also seems Deleuze uses the term 'crystal' to describe the pyramid, consciously or not, in order to highlight the debt his own 'crystal-image' owes to Leibniz's pyramid. In this mythical pyramid, as in the crystal-image, virtual and actual coalesce – a perquisite, as Deleuze suggests, for modern literature.

Leibniz chooses to end the *Theodicy* by recounting a myth concerning Sextus Tarquinius, son of the last Etruscan king ruling Rome. The Roman Republic was founded following the dethronement of Sextus after he raped Lucretia, wife of Lucius Tarquinius Collatinus; thus Sextus' fate played a foundational role in Roman history which lies at the background of Leibniz's retelling. The myth justifies Sextus' punishment, thereby supporting the existence of the Roman republic, and, indirectly, that of the Holy Roman Empire supported by Leibniz.

The first among four levels of dialogue, between two speakers, 'Laurentius' representing the author, and 'Antonio' his opposer, is freely reproduced from Laurentius Valla's *Dialogue on Free Will.*[148] As Leibniz concedes, Valla's dialogue was meant to refute what he perceived as the pagan position of Boethius in *Consolation of Philosophy*, to which Leibniz responds, as well, in the *Theodicy.*[149] In his dialogue, Valla invokes the myth of Sextus, who went to Apollo seeking a prophecy of his future, only to learn his life would be short and unhappy. Leibniz provides the final words of his position on the conflict between free will and pre-destination by intervening in this debate between Boethius and Valla.

Leibniz's retelling includes the additional figures of Theodorus, the high priest who was present during the conversation between Jupiter and Sextus, and Pallas Athena, Jupiter's daughter. The fourth and most extensive level consists of a dialogue between the two, following Theodorus' travelling to Athens and falling asleep in the Goddess's temple. He wakes up in an unknown land, before a palace. Pallas Athena, daughter of Jupiter appearing at the gates, speaks to him: 'You see here the palace of the fates, where I keep watch and ward. Here are representations not only of that which happens but also of all that which is possible.'[150] Pallas shows Theodorous a multiplicity of possible worlds, brought to life by her voice, every one of which represents a possible fate for Sextus.

Leibniz's version of the fable introduces, as mentioned, an architectural wonder: a splendid palace from without, which becomes a pyramid when viewed from within. Similarly, every hall, or chamber of the palace, turns into a world when Theodorous steps into it.[151] As in a dream, internal and external are one and the same yet distinct. Appearances are deceiving, the palace-pyramid supplying a multiple-layered experience in which true and false can no longer be distinguished from one another. Theodorous' dream resembles contemporary virtual reality experiences, activated by touch:

> Theodorous saw the whole life of Sextus as at one glance, and as in a stage presentation. There was a great volume of writings in this hall: Theodorus could not refrain from asking what that meant. It is the history of this world which we are now visiting, the Goddess told him; it is the book of its fates. You have seen a number on the forehead of Sextus. Look in this book for the place which it indicates. Theodorus looked for it, and found there the history of Sextus in a

> form more ample than the outline he had seen. Put your finger on any line you please, Pallas said to him, and you will see represented actually in all its detail that which the line broadly indicates. He obeyed, and he saw coming into view all the characteristics of a portion of the life of that Sextus.[152]

This passage is remarkable for its construction of a multi-media artistic experience including live performance, sound, vision, text and touch; a 'total installation', as Schaub writes, or a *Gesamtkunstwerk*.[153] As Benjamin mentions, the *Gesamtkunstwerk* was considered an ideal for the *Trauerspiel* form, an 'interweaving of all arts' essential to allegorical creation.[154] Leibniz's retelling of Sextus' myth is moreover allegorical *par excellence* in its abbreviation of history into spatial forms: the book of fates lies within the palace of fates, upon the pages of which Sextus' multiple histories are condensed into symbols, in yet another folded structure of expression.

It is not by chance that it is Pallas Athena, the goddess of wisdom, who guides Theodorous in his journey. Athena stands for 'Lady Philosophy' in Boethius' *Consolation of Philosophy*.[155] Here, Leibniz takes Boethius' side in the debate against Valla, defending the virtues of philosophical reasoning. As per the closing words of the *Theodicy*, the source of things must be sought in knowledge or intelligence.[156] A key argument of the *Theodicy* conditions freedom upon rational thought, hence Athena as a final guide to wisdom provides it with a fitting ending. By designating this as the myth of origin of modern literature, Deleuze highlights, then, the philosophical origins of modern literature and cinema.

When Theodorous reaches the apex of the pyramid, he enters an ecstatic state and is drugged by Athena, who places a drop of 'divine liquid' under his tongue: only in a dream and under drugs can the best world and true happiness be experienced.[157] When drugged, that is, when not in full control of his intellectual and perceptual capacities, a state less clear and translucent, Theodorous can finally experience the real world. As Leibniz writes at the end of 'Principles of Nature and Grace', happiness consists in constant striving for unattainable perfection; hence it is impossible for Thedorous to experience the world at the apex of the pyramid unaided by narcotics.

Leibniz's 'crystal pyramid' is the most complete expression of his theory of multiple compossible worlds, serving as final proof of his theory of God's selection of the best world out of an infinity of possible ones. Theodorus finally reaches the 'real true world' at the top of the pyramid, where he understands that Sextus' sin, committed freely, had served the greater purpose of founding the

Roman Empire. One of the morals of the story is that although everything has a reason, according to Leibniz's 'principle of sufficient reason', it is often hidden from mortal eyes.

For Deleuze, there is no best of possible worlds, and the immanent presence of multiple worlds within our own articulated in *The Fold* was foreshadowed by the figure of eternal return as described in *Difference and Repetition*. In his interpretation of the fable, Deleuze describes the multiplicity of worlds existing together in the palace as virtuals, actualized in the special circumstance of the dream. Leibniz's cinematic solution to the 'contingent futures' paradox uses the category of 'incompossibility', which precedes that of possibility. However, for Deleuze, in the 'crystalline' regime all 'incompossible' worlds belong to one and the same universe, with the result that 'narration … becomes fundamentally falsifying'.[158] Deleuze cannot accept Leibniz's opposition between the categories of truth and falsity, and therefore extends the mere interchangeability of true and false in Leibniz's myth, stretching Leibniz's intermediary category of 'compossibility' further than originally intended.

Benjamin does not refer to this fable nor to Leibniz's *Theodicy* more generally, yet one could imagine he would have appreciated Leibniz's multi-layered, interpolative account. Although Benjamin's definitions of modern literature include a destructive element lacking from Leibniz's moralistic myth, Leibniz's use – in the final words of the *Theodicy*, of 'total allegory' as a form of expression in which history is 'secularised in the setting' and a 'temporal process is caught up and analysed in a spatial image' – illustrates the sources of modern literary and mediatic forms in the Baroque.[159]

Conclusion

We thought there were windows but actually it's made of mirrors.
– Julieta Aranda, Brian Kuan Wood, and Anton Vidokle,
The Internet Does Not Exist

In 'In Free Fall: A Thought Experiment on Vertical Perspective', Hito Steyerl describes how our orientation in time and space has shifted in recent years due to the crumbling of the linear perspective paradigm.[1] Citing Erwin Panofsky, she describes how linear perspective is constructed around the single point of view of the immobile spectator, assuming this view to be objective. Steyerl highlights the role of the horizon, constituted by linear perspective as an abstract line, in contributing to the creation of an illusionary 'reality' viewed in a painting as through a window.

Linear perspective, she writes, began to lose its grip as the dominant paradigm in the nineteenth century, a process that was accelerated in the twentieth, culminating with the current dominance of aerial views. Our perception is shaped by a proliferation of technologies developed by the military and applied in the entertainment world, from satellites and drones to Google Earth and 3D animation. The pre-eminence of aerial representation has brought about a horizonless 'condition of free fall'. As she writes,

> If we accept the multiplication and de-linearization of horizons and perspectives, the new tools of vision may also serve to express, and even alter, the contemporary conditions of disruption and disorientation. Recent 3D animation technologies incorporate multiple perspectives, which are deliberately manipulated to create multifocal and nonlinear imagery. Cinematic space is twisted in any way imaginable, organized around heterogeneous, curved, and collaged perspectives. The tyranny of the photographic lens, cursed by the promise of its indexical relation to reality, has given way to hyperreal representations – not of space as it is, but of space as we can make it – for better or worse.[2]

The photographic image no longer prevails, making way for virtual realities more vivid than those met by Theodoros in Leibniz's crystal pyramid, invading our everyday lives and eliminating the ground beneath our feet. This disorienting state, created by a perspectival multiplicity and the contortion of space and time, redistributes power, normalizing the disembodied bird's-eye view of superiors onto inferiors. Steyerl cites Eyal Weizman, who describes the Israeli occupation of Palestine as a 'politics of verticality' in which power is distributed across a vertical line.[3]

In his conclusion to *Cinema 2*, Deleuze anticipates some of the ways in which shifts in digital technologies would reorient our perspectival bearings and bring forth a 'new regime of images':

> The new images no longer have any outside (out-of-field), any more than they are internalized in a whole; rather, they have a right side and a reverse, reversible and non-superimposable, like a power to turn back on themselves. They are the object of a perpetual reorganization, in which a new image can arise from any point whatever of the preceding image. The organization of space here loses its privileged directions, and first of all the privilege of the vertical which the position of the screen still displays, in favour of an omni-directional space which constantly varies its angles and co-ordinates, to exchange the vertical and the horizontal. And the screen itself, even if it keeps a vertical position by convention, no longer seems to refer to the human posture, like a window or a painting, but rather constitutes a table of information, an opaque surface on which are inscribed 'data', information replacing nature, and the brain-city, the third eye, replacing the eyes of nature.[4]

Rather than pronouncing the demise of the analogue photographic image, Deleuze describes a shift in the function of images, which become plastic – no longer having an 'outside', a beginning or end. They construct an 'omnidirectional space' that does not privilege a specific point of view, the reign of the vertical giving way to an exchange between vertical and horizontal. The screen constitutes a surface inscribed with data instead of purporting to represent nature.[5] It would have been interesting to hear Deleuze's thought on the domination of smart phones, those petite vertical screens that contain multiple worlds, accelerating the omni-directionality of our space even further, in contemporary life.

Leibniz was famously the inventor of a binary numerical system that served as the basis of binary code, contributing to the inception of digital technologies.[6] Alongside the calculating machine he constructed in order to implement the code, these are seen as the earliest inventions contributing to the foundation of the internet.[7] Both Leibniz's binary numerical system and the universal mathematical language he imagined, *characteristica universalis*, function according to the logic of expression, which assumes every facet of reason may be expressed in notation or numerals.[8] The very same logic underlies tables of information and some contemporary data systems. In *The Fold*, Deleuze observes that Leibniz's monads, or simple substances, are better described as 'tables of information' inscribed with coded lines, than as windows.[9]

As we have seen in Chapter 4, Leibniz engaged in research in projective geometry, which served as the foundation for linear perspective. Leibniz contributed, then, both to the development and crowning of the paradigm of linear perspective and to the logic that guides the technologies that effected its downfall and subsequent replacement by virtual realities and excess of information. While Steyerl's account, quoted earlier, suggested that de-linearization has, in the last century, triumphed over linear perspective, the shared origin of both linear and di-linear perspectives in Leibniz suggests a profound link between organizing systems and disorder, as exemplified by the internet itself.[10]

In their readings of Leibniz, Benjamin and Deleuze emphasize the disruptive logic that guides Leibniz's philosophy, beyond its harmonious façade. Their Leibniz is a philosopher of *désaccords* alongside *accords*, not the naïve optimist of Voltaire's *Candide*.[11] The ways they read Leibniz reflect the tendencies towards the catastrophic and vertiginous in their own philosophies. Deleuze's fascination with the opaque crystal fits well into Culp's reading of a 'dark Deleuze', against the tendencies of reading him as a uniquely affirmative thinker whose philosophy is devoid of destructive forces.[12]

This aspect of both Deleuze and Benjamin's readings of Leibniz paints a very different picture of Leibniz's philosophy from the renowned philosophical readings of Leibniz conducted in the early twentieth century, such as those of Russell, Couturat Cassirer, and Husserl, who appropriated his philosophy into logical, neo-Kantian or phenomenological domains.[13] Russell and Coutourat's readings were extremely influential in paving a path for Leibniz's reception in the twentieth and twenty-first centuries that privileged his logical and mathematical writings while deeming other parts of his work 'exoteric' and therefore unworthy of serious consideration.[14] Their readings remain influential today, at a time when a more comprehensive collection of Leibniz's writing than ever before is available.

The Leibniz who emerged from readings by Russell-Couturat and their followers is a systematic thinker and precursor of Kant's rational philosophy. This approach still dominates the field of Leibniz scholarship, although other voices are beginning to emerge.[15] Deleuze and Benjamin's readings open up a lesser-known Leibniz, one suitable for the critical examination of our times, just as Spinoza's philosophy was revived in post-1968 Marxist theories.[16] Theirs are conceptual, philosophical readings that view Leibniz's thought as rooted in the Baroque, yet at the same time as actual and relevant for their own times – both appropriate his concepts into their own philosophy, breathing new life into them, letting them grow and change.

Deleuze, however, stays within the 'logical Leibniz' paradigm. Notwithstanding the creativity and originality of his reading of Leibniz in *The Fold*, it is still determined by his understanding of Leibniz's concept of expression as logical rather than symbolic and his preference of an unlimited infinite over a limited one. As we have seen in Chapter 2, Benjamin criticizes Cohen for his 'logical' concept of expression. Deleuze, at heart a Spinozist who calls for univocity of being, puts forward an unlimited concept of infinity. Yet without limitation, as Benjamin points out, the infinite tends to collapse into itself in regress.[17]

In 'On Freedom' Leibniz introduces the image of the two labyrinths previously discussed, which he later replicates in the *Theodicy*, using a slightly different formulation: 'For there are two labyrinths in which the human mind is caught. One concerns the composition of the continuum; the other concerns the nature of freedom. And both arise from the same source, namely, the infinite.'[18] The source of both philosophical problems, that of continuity and that of freedom, is the concept of the infinite. Any reading of Leibniz hinges, therefore, upon its interpretation of the infinite, and for this reason, I have highlighted the divergence between Deleuze and Benjamin on this aspect of Leibniz's philosophy.

While in the present work Benjamin and Deleuze's readings of Leibniz have served to open up a wider comparison of their philosophies, these readings diverge in their interpretation of this most significant aspect of Leibniz's thought. This divergence has important implications for the ways in which they understand temporality; Deleuze's critique of Leibniz led him to privilege the unlimited infinity of eternal return, while Benjamin's emphasizes the discontinuity of history as manifest in the monadic structure of the event. As we have seen in Chapter 5, Benjamin had drawn a distinction between 'qualitative infinity' and 'infinitely empty time' already in his dissertation on the Romantics.[19] For Benjamin, as discussed in Chapter 3, the standstill or cesura in the continuum of history consists of the mirror-image of the 'state of exception' in which the sovereign's power is suspended.[20] Hence the powerful revolutionary force of the limitation of infinite movement.

Deleuze's *The Fold* is written from a position of admiration for Leibniz's philosophy, yet Deleuze's critique of Leibniz as ultimately a philosopher of identity, formulated most clearly in *Difference and Repetition*, remains in place. Leibniz's philosophy is therefore viewed from a critical distance, mediated by the cultural analysis of the Baroque.

Nevertheless, Leibniz's multiplicity of virtual worlds expressed in our own and the category of *incompossibility* were instrumental for Deleuze's definition, in *Difference and Repetition*, of the categories of singularity and difference. Kant wrote of Leibniz that 'we must not allow ourselves to be perplexed by his explanation of sensation as a confused kind of perception'.[21] Deleuze returns to Leibniz from a post-Kantian perspective in order to redefine perception through sensation. Leibniz's concepts of 'virtuality' and 'force' are viewed as the precursors of Bergson and Nietzsche's concepts, which Deleuze more adamantly adopts.

In contrast, I have showed that in both Benjamin's early and later appropriations of Leibniz, what is at stake for Benjamin is a refiguration of historical knowledge and experience. The expressive nature of the monad and the perspectival nature of the universe persistently appear in Benjamin's writing in different guises: for example, in the 'fore and subsequent life' [*Vor- und Nachleben*] of the work of art, and in his concept of 'virtual history'. Benjamin's 'historical perspectivism', through which he integrates Leibniz's concepts into his philosophy of historical experience, is significant in its contribution to reassessing Leibniz's philosophy of history, a scarcely researched aspect of his work although historiography was his chief concern in later life.[22]

In Benjamin's conception of historical experience, the monadic, expressive construction of history and the 'intensive image' constitute the 'montage' of history. That is, in Benjamin's late writing on history, expression equals construction, as acutely observed by Alliez in relation to Deleuze and Guttari's *What Is Philosophy?*[23] Expression is redefined through construction and therefore montage: a novel redefinition of expression that includes constructive and destructive elements. In his last work, Deleuze approaches Benjamin's interpretation, yet without going so far as to introduce destruction into expression.

Against readings of Leibniz as primarily a philosopher of reconciliation, following Benjamin allows a reading of Leibniz's disparate and eclectic texts as montage or constellation rather than searching for the underlying 'synthetic project'.[24] Although Deleuze's reading of Leibniz, especially in *The Fold*, is more elaborate and detailed, and based on a more comprehensive acquaintance with Leibniz's texts, it is Benjamin's reading that better succeeds in breaking the 'logical paradigm' dominating Leibniz's reception, extending the 'anti systematic impulse' within Leibniz's apparent systematicity and uncovering the political potential of Leibniz's texts.

Deleuze and Benjamin both read Leibniz's concept of freedom through Nietzsche's, finding the former as lacking in opposition to the latter. Deleuze's Nietzschean Leibniz is first and foremost a Leibniz of eternal return, in which the multiplicity of virtual worlds is already included in our own. Benjamin's may be seen as a nihilistic Nietzschean-Leibniz, a Leibniz of destruction, not only of perfection – an 'anti-Enlightenment', irrational Leibniz rather than simply a 'pre-Enlightenment' savant. Benjamin's could be termed a deconstructive reading, or one that reads Leibniz against Leibniz.[25]

Deleuze and Benjamin's readings converge in their redefinition of perception through bodily sensation against Kant's theory of experience, resulting in corporeal pedagogies that view learning as involving the body. Leibniz's bodies never cease from movement, and Benjamin and Deleuze draw on his conceptions of force, movement, life – which explains why Leibniz's philosophy is crucial for their understanding of how our perception is shaped by cinema, the art of moving images. Deleuze consciously developed this aspect of Leibniz's philosophy, as he states, 'it's organisms that die, not life. Any work of art points a way through for life, finds a way through the cracks. Everything I've written is vitalistic, at least I hope it is'.[26] However, while Deleuze's appropriations of Leibniz's concepts of force and life result in a pedagogy, Benjamin's results in a politics. This is manifest for example in Benjamin's description of life forces motivating society as a body that materialize into 'revolutionary discharge' in the Surrealism essay.[27]

Benjamin, unlike Deleuze, had a complex relation with vitalism. On the one hand, as already noted, his concept of experience [*Erfahrung*] contained a critique of 'lived experience' [*Erlebnis*], tinged with criticism of the life philosophy [*Lebensphilosophie*] that was in vogue in his time and whose representatives included Bergson, Nietzsche and Simmel. This critique was explicitly stated, as narrated by Scholem, in Benjamin's adamant rejection of Buber's 'cult of lived-experience [*Erlebnis*]'. Benjamin was critical, as Jay observes, of the manner in which the concept of *Erlebnis* was used to commend the First World War.[28] On the other hand, beyond the significance of corporeal experience for Benjamin, terms such as 'life' and 'energy' abound throughout his writing. McLaughlin argues that Benjamin formulated what he terms a 'biophilology', beginning in his early 'The Life of Students' (1914–15) and running through to his texts on Baudelaire written in the late 1930s.[29] Benjamin's position vis-à-vis Bergson and Nietzsche's philosophies was both admiring and critical, which further complicates the picture. The tendency of critics to obscure or deride vitalistic elements in Benjamin's philosophy, shaped by Adorno's early critique of these aspects, makes it even more difficult to properly evaluate the function of these conceptions in Benjamin's writing.[30]

Chapter 1 introduced Leibniz's concepts of expression and monad and traced the ways these were appropriated by Benjamin and Deleuze in their earlier works. For both authors, we have seen, the monad offers structures of immanence and inclusion which correspond to their interpretations of the Baroque. Chapter 2 focused on Benjamin and Deleuze's theories of learning and how these draw on Leibniz's theory of 'minute perceptions' as part of their critiques of Kant's epistemology. Both authors also reformulate Plato's theory of learning as remembrance, while challenging his conception of innate knowledge.

In Chapter 3, I have shown how Benjamin's philosophy of history was informed by Leibniz's conceptions of force and virtuality. The recurrent description of force as self-incurred and the theme of automatization in Benjamin's writing from different periods make manifest that not only the debate around *Lebensphilosophie* but also Leibniz's understanding of substance as self-moving and intensive played a role in Benjamin's 'vitalism', which, in the 'Artwork' essay and essays on Baudelaire, for example, combined with his response to nascent forms of technology, as discussed in Chapters 4 and 5. We have also shown in these chapters how Benjamin's 'historical perspectivism' and Deleuze's 'Mannerist perspectivism' draw on their concepts of allegory and Mannerism, respectively, and how these contribute to their concepts of the 'crystal' and 'dialectical' image and montage.

Deleuze noted Benjamin's preoccupation with automatic movement in the Artwork essay:

> Walter Benjamin's article set itself inside cinema in order to show how the art of automatic movement (or, as he ambiguously said, the art of reproduction) was itself to coincide with the automization of the masses, state direction, politics become 'art': Hitler as film-maker[31]

Deleuze reads Benjamin as arguing for a link between technological reproduction and the automatism of the collective as body with the consequence of the 'aestheticization of politics'.[32] As we have seen, Benjamin brings the two together when describing the 'collective physis' in the essay on Surrealism. Automatism is significant for Deleuze's own understanding of cinema; as he writes, 'if cinema is automatism become spiritual art – that is, initially movement-image – it confronts automata, not accidentally, but fundamentally'.[33] Cinema is at its most essential level automated movement, like that which drives Leibniz's 'entelechies' or self-moving monads.

Benjamin and Deleuze's fascination with 'becoming molecular' (Deleuze) or 'miniaturisation' (Benjamin), often linked, in their writing, with the automated production of cinematic or photographic images, also has its sources in Leibniz's logic of expression. Johannßen argues that Benjamin anticipated the shift from mechanical reproduction to digitization in his theory of reproducibility.[34] This was possible, he suggests, by Benjamin's development of a 'counter concept of miniaturisation' based on the expressive nature of Leibniz's theory of 'minute perceptions', in which details receive close attention.[35] Better suited to describe digital technology, in a sense Leibniz's concept of expression enables a leap over the analogue phase, directly to the horizonless condition described by Steyerl.

Although Leibniz's understanding of time as a relation of succession may have contributed to the development of progressive visions of historiography, his notions of expression and perspectivism, developed within his metaphysics and based on his mathematics and geometry, enable the reimagining of multi-perspectival space and time that travels in circuits and triangles against linear time and space, paving the way for the trail Deleuze blazes between chaos and the brain.[36] The way to the chaotic perspectivism of the internet, which Deleuze anticipates in his writing, traverses through Benjamin's historical perspectivism. Deleuze, but even more so Benjamin, show us the ways in which we are, and will continue to be, 'still Leibnizian'.[37]

Notes

Introduction

1 Gilles Deleuze, *Cours Vincennes*, 20/01/1987, https://www.webdeleuze.com/textes/135.
2 Alain Badiou, *Deleuze: The Clamor of Being*, trans. Louise Burchill (Minneapolis, mn: University Of Minnesota Press, 2000), 1.
3 Deleuze. 'The New Harmony', in *The Fold: Leibniz and the Baroque* (London: The Athlone Press, 1993), 121–39.
4 Leibniz, 'Pacidus Philalethi', in *Opuscules et fragments inédit de Leibniz*, ed. Louis Couturat (Paris: Félix Alcan, 1903), 614–15, quoted in Deleuze, *The Fold*, 6.
5 See for example Gottfried Wilhelm Leibniz, *Theodicy: Essays on the Goodness of God, the Freedom of Man, and the Origin of Evil*, trans. Austin Farrer, ed. E. M Huggard (New York: Cosimo, 2009), 68, 69.
6 See Walter Benjamin, *The Correspondence of Walter Benjamin: 1910–1940*, ed. Gershom Scholem, Paperback ed (Chicago, Ill London: Univ. of Chicago Press, 2012), 260–6.
7 Walter Benjamin, *Origin of the German Trauerspiel*, trans. Howard Eiland (Cambridge, MA: Harvard University Press, 2019).
8 Benjamin, ibid, 41.
9 Dosse describes a course Deleuze taught during his studies at the *classes préparatoires littéraires*, in 1955–6 on the question of foundations in philosophy, which focused on Leibniz and Heidegger. See François Dosse, *Gilles Deleuze et Félix Guattari: biographie croisée* (Paris: La Découverte, 2009), 132.
10 Gilles Deleuze, *Expressionism in Philosophy: Spinoza* (New York: Cambridge, MA: Zone Books; Distributed by MIT Press, 1990); Gilles Deleuze and Paul Patton, *Difference and Repetition* (London: Continuum, 2004); Gilles Deleuze et al., *The Logic of Sense* (New York: Columbia University Press, 1990).
11 Benjamin, 'The Concept of Criticism in German Romanticism', in *Selected Writings, Volume 1: 1913–1926*, ed. Marcus Bullock and Michael W. Jennings (Cambridge; London: Belknap Press, 2004), 116–200.
12 Gottfried Wilhelm Leibniz, 'The Principles of Philosophy, or, the Monadology (1714)' and 'Discourse on Metaphysics (1686)', in *Philosophical Essays*, ed. Roger Ariew, and Daniel Garber (Indianapolis: Hackett Pub. Co, 1989). I will also use the versions in Gottfried Wilhelm Leibniz, *The Monadology and Other Philosophical Writings*, ed. Robert Latta (London: Oxford University Press, 1971) and Gottfried

Wilhelm Leibniz, *Leibniz's Monadology: A New Translation and Guide*, ed. Lloyd Strickland (Edinburgh: Edinburgh University Press, 2014) as well as Gottfried Wilhelm Leibniz *La Monadologie*, édition électronique, édition annotée par Émile Boutroux, initialement publiée chez C. Delagrave, Paris, 1881. See Benjamin, *Origin of the German Trauerspiel*, 27 and fn. 36; Walter Benjamin, *Gesammelte Schriften. Bd. 1 Teil 1: Abhandlungen*, ed. Rolf Tiedemann and Hermann Schweppenhäuser, 9. Auflage, Suhrkamp-Taschenbuch Wissenschaft 931 (Frankfurt am Main: Suhrkamp, 2019), 228. See also Peter Fenves, who mentions that Benjamin refers to Leibniz's monads in the 'Discourse' although Leibniz did not use the term 'monad' in this text, only in texts written at a later stage. *The Messianic Reduction: Walter Benjamin and the Shape of Time* (Stanford, CA: Stanford University Press, 2011); fn. 21, 281.

13 The list was compiled between 1917 and 1920. See Walter Benjamin-Archiv, Akademie der Künste, Berlin, MS 506 varia.

14 Two recently published volumes focus on Benjamin's practices of reading and appropriation: *Entwendungen: Walter Benjamin und seine Quellen*, ed. Jessica Nietzsche and Nadine Werner (Paderborn: Wilhelm Fink, 2018) and *Material und Begriff: Arbeitsverfahren und theoretische Beziehungen Walter Benjamins*, Deutsche Originalausgabe, Argument Sonderband, neue Folge, AS 322 (Hamburg: Argument, 2019).

15 Benjamin, 'Verzeichnis der gelesenen Schriften', *Gesammelte Schriften 7*, ed. Rolf Tiedemann and Hermann Schweppenhauser (Frankfurt am Main: Suhrkamp Verlag, 1972–1989), 443. Heinz Heimsoeth, 'Leibniz' Weltanschauung als Ursprung seiner Gedankenwelt: Zum 200.Todestage des Denkers am 14 November 1916', *Kantstudien* (1917), 365–95. Hermann Cohen, *Kants Theorie der Erfahrung* (Saarbrücken: VDM, Müller, 2006).

16 See Schwebel, *Walter Benjamin's Monadology* (Doctoral Thesis, Department of Philosophy, University of Toronto, 2012), 25.

17 Cassirer's *Die Begriffsform im mythischen Denken* (Braunschweig: Vieweg and Teubner Verlag, 1922) appears in Benjamin's list of books he had read, Benjamin, Walter, *GS*7, 452.

18 Ferrari, Massimo. *Ernst Cassirer: dalla scuola di Marburgo alla filosofia della cultura*, ed. Voight et al. (Firenze: L.S.Olschki, 1996) 45–100 and 171–190, quoted in Tagliacozzo, Tamara. *Experience and Infinite Task: Knowledge, Language, and Messianism in the Philosophy of Walter Benjamin*. Founding Critical Theory. London; New York: Rowman & Littlefield International, 2018, 101.

19 See Kevin Mclaughlin, 'Virtual Paris: Benjamin's Arcades Project', in *Benjamin's Ghosts: Interventions in Contemporary Literary and Cultural Theory*, ed. Gerhard Richter (Stanford, Calif: Stanford University Press, 2002), fn23, 342. See Richard Herbertz, *Die Lehre vom Unbewußten im System von Leibniz* (Niemeyer: Halle, 1905).

20 Such editions include, for example, G.W. Leibniz, *Die philosophischen Schriften*, ed. C.J. Gerhardt (Berlin, 1879–1890) and G.W. Leibniz, *Nouvelles lettres et opuscules inédits de Leibniz*, ed. Foucher de Careil (Paris, 1857–75). As Sjoerd van

Tuinen suggests, Deleuze did not necessarily use the latest available editions of Leibniz's work. See *Mannerism in Philosophy* (Doctoral Thesis, Faculty of Arts and Philosophy, University of Ghent 2009), introduction, 7.

21 Salomon Maimon, *Versuch über die Transzendentalphilosophie*, ed. Florian Ehrensperger Philosophische Bibliothek 552 (Hamburg: Meiner, 2004).

22 For studies on Deleuze's reception of Leibniz see: Alex Tissandier, *Affirming Divergence: Deleuze's Reading of Leibniz* (Edinburgh: Edinburgh University Press, 2022); Sjoerd van Tuinen and Niamh McDonnell, eds., *Deleuze and The Fold: A Critical Reader* (Basingstoke: Palgrave Macmillan, 2010); Alain Badiou, 'Gilles Deleuze, The Fold: Leibniz and the Baroque', in *Gilles Deleuze and the Theater of Philosophy* (London: Routledge, 2019), 51–69; Guillaume Sibertin-blanc, 'Pli psychotique et maniérisme, sur le leibnizianisme clinique de gilles deleuze', in A. Jdey (dir.) *Gilles Deleuze, la logique du sensible – Esthétique & clinique* (Paris: Editions de l'incidence, 2013), 357–74; Daniel W. Smith, 'Deleuze on Leibniz: Difference, Continuity, and the Calculus', in *Essays on Deleuze* (Edinburgh: Edinburgh University Press, 2012), 43–58; Lang Baker, 'The Cry of the Identicals: The Problem of Inclusion in Deleuze's Reading of Leibniz', *Philosophy Today* 39, no. 2 (Summer 1995): 198–211. For studies on Benjamin's reception of Leibniz see: Peter D. Fenves, *Arresting Language: From Leibniz to Benjamin* (Stanford, CA: Stanford University Press, 2001); Peter D. Fenves, *The Messianic Reduction: Walter Benjamin and the Shape of Time*, Samuel Weber, *Benjamin's -abilities* (Cambridge, MA: Harvard University Press, 2010); Paula Schwebel, *Walter Benjamin's Monadology*, 'Monad and Time, Reading Leibniz with Benjamin and Heidegger', in *Sparks Will Fly: Benjamin and Heidegger*, ed. Andrew Benjamin and Dimitris Vardoulakis (Albany: State University of New York Press, 2015), 123–44; Paula Schwebel, 'Constellation and Expression in Benjamin and Leibniz', in *Thinking in Constellations: Walter Benjamin and the Humanities*, ed. Caroline Sauter and Nassima Saraoui (Cambridge: Cambridge Scholars Publishing, 2017), 51–82; Fabrizio Desideri, 'Intermittency: The Differential of Time and the Integral of Space, The Intensive Spatiality of the Monad, the *Apokatastasis* and the Messianic World in Benjamin's Latest Thinking', *Aisthesis rivista on-line del Seminario Permanente di Estetica* IX, no. 1 (2016): 177–87; Dennis Johannßen, 'Gottfried Wilhelm Leibniz. Andacht zum Kleinen: zu Benjamins Leibniz-Lektüren', in *Entwendungen: Walter Benjamin und seine Quellen*, ed. Nadine Werner and Jessica Nietzsche (Paderborn: Wilhelm Fink, 2018), 281–302; Ilit Ferber, 'Melancholy and Truth', in *Philosophy and Melancholy: Benjamin's Early Reflections on Theater and Language*, Cultural Memory in the Present (Stanford, California: Stanford University Press, 2013).

23 Several studies that examine the philosophies of Benjamin and Deleuze together are the following: Timothy Flanagan, *Baroque Naturalism in Benjamin and Deleuze: The Art of Least Distances* (Cham, Switzerland: Palgrave Macmillan, 2021), 'The Free and Indeterminate Accord of "The New Harmony": The

Significance of Benjamin's Study of the Baroque for Deleuze', in *Deleuze and The Fold: A Critical Reader* (Basingstoke, UK; New York, NY: Palgrave Macmillan, 2010), 46–64; Olaf Berg, 'Benjamin und Deleuze: Ansätze für eine kritische Geschichtswissenschaft in Filmbildern', *Zeitschrift für kritische Theorie* 12 (2006): 68–97; Peter Szendi, *Le supermarché du visible* (Paris: Minuit, 2017); Timothy Edward Flanagan, 'Presentation of Reciprocity: An Interpolative Study of the Baroque Works of Walter Benjamin and Gilles Deleuze', PhD diss., University of Dundee, 2007. However, none of the above interrogates the role of Leibniz's philosophy and mathematics as a crucial intermediary between Deleuze and Benjamin.

24 Paula Schwebel's dissertation, *Walter Benjamin's Monadology*, and her book chapters 'Monad and Time, Reading Leibniz with Benjamin and Heidegger' and 'Constellation and Expression in Benjamin and Leibniz' comprise an extensive reading of Benjamin's early reception of Leibniz. One of the rare readings of Benjamin's later references to Leibniz is conducted by Desideri in 'Intermittency: The Differential of Time and the Integral of Space, The Intensive Spatiality of the Monad'.

25 Lang Baker's 'The Cry of the Identicals', in which the author attempts to understand the shifts in Deleuze's reception of Leibniz, is an exception.

26 Manuel DeLanda highlights, for example, the significance of isomorphism as that which underlies the structures of strata and meshwork in *A Thousand Plateaus* with no mention of Leibniz's expression as a key source for Deleuze's attention to isomorphic structures. See 'Immanence and Transcendence in the Genesis of Form', in *A Deleuzian Century?*, ed. Ian Buchman (Durham, NC: Duke University Press, 1999), 281–317.

27 See for example Bowden, Sean and Clisby, Dale, eds. 'The Virtual, the Actual, and the Intensive Contentions, Reflections and Interpretations'. *Deleuze Studies*, 11, no. 2, 2017. Despite the terms listed in the title, none of the articles in this issue discuss Deleuze's reception of Leibniz.

28 As Benjamin writes, 'If this redeemed being is ascertained to be in the idea, then the presence of the natural-historical fore- and after-history – as opposed to history proper – is virtual … The deepening of historical perspective in such investigations, whether into the past or into the future, in principle knows no bounds'. Benjamin, *Origin of the German Trauerspiel*, 27. Dipesh Chakrabarty develops Benjamin's concept of 'deep history' in relation to concepts of the Anthropocene and geological time in 'Decentring the Human?', *The Human Condition in the Anthropocene*, The Tanner Lectures in Human Values, Delivered at Yale University, February 18–19, 2015, 181.

29 Ibid.

30 Benjamin, 'Experience and Poverty', *SW*2, 731–2.

31 See for example a letter to Scholem from 22 December 1924, in which Benjamin expresses surprise at his 'various points of contact' with 'radical Bolshevist theory' *The Correspondences*, 258.

32 See Scholem's letter to Benjamin from 20 February 1930. In the background of Scholem's confrontation is a university position he had arranged for Benjamin in Jerusalem upon the latter's promise that he would study Hebrew and come for a prolonged stay in the city, a promise Benjamin was unable to keep. In Benjamin's reply he writes that 'the question of my relationship to Judaism is always the question of how I stand … in relation to the force you have touched in me'. Letter from 25 April 1930, *The Correspondences*, 364.

33 'Theodor W. Adorno to Walter Benjamin' letter from 10 November 1938. *The Correspondences*, 582.

34 Scholem recognized this eventually, writing in his memoir that 'there now appeared an often puzzling juxtaposition of the two modes of thought, the metaphysical-theological and the materialistic, so that the two became intertwined. This interlocking of two elements that by nature are incapable of balance lends precisely to those of Benjamin's works that derive from this attitude their significant effect and that profound brilliance that distinguishes them so impressively from most products of materialistic thought and literary criticism, noted for their uncommon dullness'. See Scholem, Gershom. *Walter Benjamin: The Story of a Friendship*. New York Review Books Classics. New York: New York Review Books, 2001, 244.

35 See Gilles Deleuze and Claire Parnet, 'E as in Enfance', in *Gilles Deleuze from A to Z*, Semiotext(e) foreign agents series (Cambridge, MA: Distributed by The MIT Press, 2011).

36 Gilles Deleuze, 'On Philosophy', in *Negotiations: 1972–1990* (New York: Columbia University Press, 1995), 137.

37 John Sellars. 'Gilles Deleuze and the History of Philosophy'. *British Journal for the History of Philosophy* 15, no. 3 (2007): 556.

38 Ibid, 139–40. See also Deleuze's statement that 'what I'm interested in are the relations between the arts, science and philosophy. There's no order of priority among these disciplines. Each is creative. The true object of science is to create functions, the true object of art is to create sensory aggregates and the object of philosophy is to create concepts'. *Negotiations*, 123.

39 Ibid, 139.

40 Elisabeth Roudinesco, *Philosophy in Turbulent Times: Canguilhem, Sartre, Foucault, Althusser, Deleuze, Derrida*, paperback ed (New York Chichester: Columbia University Press, 2010), 135.

41 Ibid, xii.

42 See Balibar, Étienne. Structuralism: A Destitution of the Subject?, trans. J. Swenson. *Differences: A Journal of Feminist Cultural Studies* 14, no. 1

(2003): 1–21; Campbell, Iain. 'Structuralist heroes and points of heresy: recognizing Gilles Deleuze's (anti-)structuralism', *Continental Philosophy Review*, 2021. https://doi.org/10.1007/s11007-021-09562-9, Craig Lundy, From Structuralism to Poststructuralism, *The Edinburgh Companion to Poststructuralism*, ed. Benoît Dillet, Iain Mackenzie, and Robert Porter (Edinburgh: Edinburgh University Press, 2013), 69–92.

43 Deleuze, 'How Do We Recognize Structuralism?', in *Desert Islands and Other Texts, 1953–1974*, ed. David Lapoujade, translated by Mike Taormina Semiotext(e) Foreign Agents Series (Los Angeles, CA: Cambridge, MA: Semiotext(e); Distributed by MIT Press, 2004).

44 See Lucie Kim-Chi Mercier, 'Michel Serres's Leibnizian Structuralism', *Angelaki* 24, no. 6 (2019): 3–21.

45 As he writes, 'the best inventors of the Baroque, the best commentators have had their doubts about the consistency of the notion, and have been bewildered by the arbitrary extension that, despite themselves, the notion risked taking. The Baroque was seen as being restricted to one genre (architecture) or to an increasingly restrictive determination of periods and places, or yet again to a radical avowal: the baroque never existed. It is nonetheless strange to deny the existence of the baroque as we deny that of unicorns or pink elephants'. Deleuze, *The Fold*, 33, translation modified.

46 As Gregg Lambert suggests in *On the (New) Baroque*, Rev. ed., Critical Studies in the Humanities (Aurora, Colo., USA: Davies Group, 2008), where he discusses several waves of Baroque literature, such as its return in the postmodern writing of Michel Foucault and Jacques Derrida, and in the Latin American *neobarroco.* See also Timothy Murray's *Digital Baroque: New Media Art and Cinematic Folds* (Minneapolis: University of Minnesota Press, 2008), in which the author shows how the Baroque fold lives on in cinema and new media.

47 Benjamin, *Origin of the German Trauerspiel*, 24.

48 Ibid, 38.

49 See Helen Hills, 'The Grit in the Oyster of Art History', in *Rethinking the Baroque,* ed. Helen Hills (Surrey; Burlington, VT: ASHGATE, 2011), 11–38.

50 Benjamin, ibid, 50.

51 Deleuze, *The Fold*, 68.

52 Ibid.

53 See Louis Couturat, 'Sur la Métaphysique de Leibniz (avec un opuscule inédit)', *Revue de Métaphysique et de Morale* 10, no. 1 (Janvier 1902): 1–25 Presses Universitaires de France, Bertrand Russell, *A Critical Exposition of the Philosophy of Leibniz: With an Appendix of Leading Passages* (Nottingham: Spokesman, 2008), Yvon Belaval, *Leibniz critique de Descartes*, Reproduit, Collection Tel 28 (Paris: Gallimard, 1978), Martial Guéroult, *Leibniz: dynamique et métaphysique suivi d'une note sur le principe de la moindre action chez Maupertuis*, Nouv. éd. Analyse

et raisons 7 (Paris: Aubier-Montaigne, 1977), and Michel Serres, *Le système de Leibniz et ses modèles mathématiques: Etoiles, schémas, points*, 3. éd, Épiméthée (Paris: Presses Univ. de France, 1990).

54 See for example David Rabouin, quoted in François Dosse, *Gilles Deleuze et Félix Guattari: Biographie Croisée* (Paris: Découverte, 2007), 532.

55 Deleuze, 'On Philosophy', *Negotiations*, 155 and 'On Leibniz', *Negotiations*, 160–1.

56 Jane O. Newman, 'Preface', in *Benjamin's Library: Modernity, Nation, and the Baroque*, Signale: Modern German Letters, Cultures, and Thought (Ithaca, N.Y: Cornell University Press: Cornell University Library, 2011), x–xi.

57 Ibid, xiii.

58 Heinrich Wölfflin, *Renaissance und Barock: Eine Untersuchung über Wesen und Entstehung des Barockstils in Italien* (Bruckmann: München, 1888).

59 Deleuze, *The Fold*, 4–5.

60 Benjamin, Letter to Fritz Radt quoted in Thomas Y. Levin, 'Walter Benjamin and the Theory of Art History', *October* 47 (Winter, 1988): 79. See Walter Benjamin, 'Rigorous Study of Art', trans. by Thomas Y. Levin, *October*, 47 (Winter, 1988), ibid, 84–90.

61 See Benjamin, ibid, fn.22, 89.

62 Alois Riegl, *Late Roman Art Industry*, trans. Rolf Winkes, Archaeologica 36 (Roma: G. Bretschneider, 1985). As Benjamin writes, 'already in *Late Roman Art Industry* the marginal case – and indeed goldsmithery which considers itself art is just that – proved to be the point of departure for the most significant overcoming of conventional universal history with its so-called "high points" and "periods of decline."' Ibid. See Gilles Deleuze and Félix Guattari, *A Thousand Plateaus: Capitalism and Schizophrenia* (Minneapolis: University of Minnesota Press, 1987), 492–3. Erwin Panofsky, La vie et l'art d'Albrecht Dürer (Vanves: Éditions Hazan, 2012) Panofsky, Erwin; Saxl, Fritz Dürers 'Melencolia I': eine quellen- und typengeschichtliche Untersuchung, Teubner, 1923. Deleuze refers to this text in *The Fold*, fn. 8, 147.

63 Deleuze, *The Fold*, 38–9.

64 'Origins' in the sense defined by Benjamin in *Origin of the German Trauerspiel*, as we will discuss in Chapter 2.

65 Shane Denson and Julia Leyda, eds., *Post-Cinema: Theorizing 21st-Century Film* (Falmer: REFRAME Books, 2016), 1. See also 'The Persistence of Cinema in a Post-Cinematic Age', Francesco Casetti, *The Lumière Galaxy: Seven Key Words for the Cinema to Come*, Film and Culture (New York: Columbia University Press, 2015), 203–15 and Special Issue Reconsidering cameraless photography. ed. Tomáš Dvořák, *Philosophy of Photography* 13 (1), 2022 for attempts to reconsider cinema and photography in light of expanded technologies in relation to their historical origins.

66 Deleuze, 'On Leibniz', *Negotiations*, 157.

67 Voltaire, *Candide* (London: Penguin, 2001).

68 Aristotle describes perception as a reception of form without matter: 'A general point to be grasped is that each sense (aisthêtikon) receives the perceptible forms without the matter. Wax, for instance, receives the design on a signet ring without the iron or gold; it acquires the design in the gold or bronze, but not insofar as (qua) the design is gold or bronze', Aristotle, *The Complete Works of Aristotle: The Revised Oxford Translation. 1* (Princeton, NJ: Princeton University Press, 1995), 'On The Soul', Book 2, 424a17 42–3.

69 As Kant writes, '… everything intuited in space or in time, hence all objects of an experience possible for us, are nothing but appearances, i.e., mere representations, which, as they are represented, as extended beings or series of alterations, have outside our thoughts no existence grounded in itself'. Immanuel Kant et al., *Critique of Pure Reason*, 15. print, The Cambridge Edition of the Works of Immanuel Kant, general ed. Paul Guyer and Allen W. Wood (Cambridge: Cambridge University Press, 2009), A491/B519, 511.

70 As he writes in the *Monadology*, 'indeed, each Monad must be different from every other. For in nature, there are never two beings which are perfectly alike and in which it is not possible to find an internal difference, or at least a difference founded upon an intrinsic quality [*denomination*]', Leibniz, *The Monadology and Other Philosophical Writings*, 9, 222.

71 Ibid, 13, 223, modified.

72 Benjamin, *Origin of the German Trauerspiel*, 2/*GS1*, 207. Deleuze, *Difference and Repetition*, xxi.

73 Deleuze, Gilles, and Félix Guattari. *What Is Philosophy?* Translated by Hugh Tomlinson and Graham Burchell. New York: Columbia University Press, 1994, 33–4.

74 Deleuze, *The Fold*, 38–9/*Le pli: Leibniz et le Baroque*. Nachdr. Collection critique. Paris: Éd. de Minuit, 2005, 27.

75 Benjamin, *Origin of the German Trauerspiel*, 47 modified/*GS*1, 228.

76 See Walter Benjamin, *The Arcades Project*, ed. Rolf Tiedemann, trans. Eiland Howard and McLaughlin Kevin (Cambridge, MA: Belknap Press, 1999), [N1a, 1], 459.

77 Gottfried Wilhelm Leibniz, *New Essays on Human Understanding*, trans. Peter Remnant and Jonathan Bennet (Cambridge: Cambridge University Press, 1996), Preface, 57.

78 Samuel Weber, 'Genealogy of Modernity: History, Myth and Allegory in Benjamin's Origin of the German Mourning Play', *MLN* 106, no. 3, German Issue (April 1991): 472.

79 Benjamin, *Arcades Project*, [N7a, 2], 470.

80 Benjamin, Walter. 'On the Concept of History' in *Selected Writings*. Volume 4: 1938–1940 / Ed. by Howard Eiland and Michael W. Jennings. 1st pbk. ed. Cambridge, MA: Belknap Press of Harvard University Press, 2006.

81 Ibid, Thesis XVII, 396 (Benjamin's emphasis).
82 The description of time as contained in the interior of understanding recalls Kant's understanding of time as the interior form of our understanding, yet the seed metaphor implies internal genesis and actualization that run counter to Kant's view of time.
83 Ibid.
84 'The construction of the idea, as stamped by totality in contrast to its own inalienable isolation, is monadological', writes Benjamin in the Foreword to *Origin of the German Trauerspiel*, 27. Of Brecht's epic theatre he writes that it 'proceeds by fits and starts, in a manner comparable to images on a film strip'. See Walter Benjamin, 'What Is Epic Theatre' [second version] in *Understanding Brecht,* trans. Anna Bostock (London and New York: Verso 1998), 23.
85 As Deleuze writes, 'Leibniz discovers in the clear, finite, idea, the restlessness of the infinitely small, a restlessness also made up of intoxication, giddiness, evanescence and even death.' Deleuze, *Difference and Repetition*, 45.
86 For example in Leibniz, *Theodicy*, 68, 69.
87 See Christia Mercer, *Leibniz's Metaphysics: Its Origins and Development* (Cambridge; New York: Cambridge University Press, 2001), 2.
88 Bernard Le Bovier de Fontenelle, 'Eloge de M. Leibnitz', in *Histoire de l'Academie Royale des Sciences. 1716* (Paris: Impr. Royale, 1718), 94–128, quoted in Antognazza, Maria Rosa. *Leibniz: An Intellectual Biography*. 1. paperback ed. Cambridge: Cambridge University Press, 2011.
89 Leibniz, *Monadology*, 56, in *The Monadology and Other Philosophical Writings*, 248. See Omar Kholeif, ed., *Art in the Age of Anxiety* (Cambridge: The MIT Press, 2020).

Chapter 1

1 Deleuze, *Cours Vincennes*, 15/4/1980: https://www.webdeleuze.com/textes/48.
2 Leibniz, *Theodicy*, 54.
3 See Paul Rateau, *La question du mal chez Leibniz: fondements et élaboration de la Théodicée*, Travaux de philosophie 15 (Paris: Champion, 2008).
4 Rutherford, Donald. *Leibniz and the Rational Order of Nature*. 1. paperback ed. Cambridge: Cambridge University Press, 1998. See Russell, *A Critical Exposition of the Philosophy of Leibniz*; Louis Couturat, *La logique de Leibniz d'après des documents inédits* (Paris: Ancienne Librairie Germer Baillière, 1901).
5 Leibniz, *Theodicy*, 8, 128.
6 See ibid, 'I call "World" the whole succession and the whole agglomeration of all existent things, lest it be said that several worlds could have existed in different times and different places. For they must needs be reckoned all together as one

world or, if you will, as one Universe. And even though one should fill all times and all places, it still remains true that one might have filled them in innumerable ways, and that there is an infinitude of possible worlds among which God must needs have chosen the best, since he does nothing without acting in accordance with supreme reason.'

7 Leibniz, *Theodicy*, 237.

8 Leibniz, 'Primary Truths', *Philosophical Essays*, 30.

9 Leibniz, 'On Freedom', *Philosophical Essays*, 98.

10 Leibniz, 'Primary Truths', 32. The 'containment' of predicates by the complete individual means that for Leibniz, all propositions are analytic. In analytic propositions one of the two terms is included in the concept of the other, as Deleuze explains in a lecture ('Sur Leibniz' given on 20th May 1980 at Vincennes – St Denis). A well-known example for an analytic propositions is 'all triangles have three sides'. Kant, as opposed to Leibniz, distinguished between analytic propositions, such as this one, and synthetic proposition, such as 'all triangles are equilateral' – a distinction Leibniz would not have accepted. For Leibniz, the concept of the triangle contains predicates such as 'equilateral' either necessarily or contingently.

11 Deleuze, *The Fold*, 41.

12 See Richard T. W. Arthur, 'The Labyrinth of the Continuum', in *The Oxford Handbook of Leibniz*, ed. Maria Rosa Antognazza (Oxford: Oxford University Press, 2018), 275.

13 See Serres, *Le Système de Leibniz*, 216.

14 From *Nouvelles de la République des Lettres* (1687), Gottfried Wilhelm Leibniz, 'extrait d'une Lettre de M. L. sur un Principe général, utile à l'explication des loix de la nature, par la consideration de la sagesse Divine; pour servir de réplique à la réponse du R. P. M, p.351', quoted in Larry M. Jorgensen, 'The Principle of Continuity and Leibniz's Theory of Consciousness', *Journal of the History of Philosophy* 47, no. 2 (2009): 224–5.

15 As Loemker notes, Leibniz's interest in psychology dates from as early as his student years in 1663–64, when he sketched out an outline of the mind. *Philosophical Papers and Letters*, 61.

16 Leibniz, *Monadology* 1, 217.

17 See Leroy E. Loemeker, 'Introduction: Leibniz as Philosopher', in *Philosophical Papers and Letters*, 17. The conceptions of 'monad' and 'individual' play an important role in Deleuze and Benjamin's readings, and as such my reading also focuses on Leibniz's later work.

18 Leibniz, 'Principles of Nature and Grace', *Monadology and Other Philosophical Writings*, 412.

19 As Leibniz describes the natural series, 'all the orders of natural beings must necessarily form only one chain, in which the different classes, like so many links, are so closely connected with one another that it is impossible for sense or imagination to determine exactly the point where any one of them begins or ends; all the species which border or occupy so to speak disputable territory [*régions d'inflexion et de rebroussement*] being necessarily ambiguous and endowed with characteristics which may equally well be ascribed to neighbouring species'. From a letter quoted by Leibniz's biographer Guhrauer in G. W. F Von Leibniz, *Eine Biographie*, vol. 1, 'Anmerkungen', 32, requoted in Latta, ed., *The Monadology and Other Philosophical Writings*, 38.

20 'In my opinion corporeal substance consists in something quite other than being extended and occupying a place: we must, in fact, ask ourselves what it is that occupies the place', Leibniz, Epistolaad Schulenburgium (1698) (G. Math. vii. 242, quoted in *The Monadology and Other Philosophical Writings*, 28). See Descartes, 'all the variety of matter, or the diversity of its forms depends on motion', René Descartes, Valentine Rodger Miller, and Reese P. Miller, *Principles of Philosophy*, Synthese Historical Library: Texts and Studies in the History of Logic and Philosophy, v. 24 (Dordrecht, Holland; Boston, U.S.A.: Hingham, Mass: Reidel ; Distributed by Kluwer Boston, 1983) part 2, 23.

21 Anne Lise Rey discusses the link between Leibniz's developing conception of the individual and developments in his dynamics; see Anne Lise Rey, 'L'ambivalence de la notion d'action dans la Dynamique de Leibniz. La correspondance entre Leibniz et De Volder', *Studia Leibnitiana* 41, no. 1 (2009): 47–66.

22 For an analysis of the importance of Leibniz' critique of the Cartesian mechanistic method for the development of Leibniz's own ideas on force and movement, see Yvon Belaval, 'Le fondements de la physique', in *Leibniz critique de Descartes*, Reproduit, Collection Tel 28 (Paris: Gallimard, 1978), 401–2.

23 See Rey, 'L'ambivalence de la notion d'action', 48.

24 'The importance of the Leibnizian formula *mv2* is due to its introducing a potential into the state of affairs', write Deleuze and Guattari in *What Is Philosophy?*, 153.

25 Esp. *Monadology*, 12–14, *The Monadology and Other Philosophical Writings*, 223–4.

26 See Leibniz, letter to Des Bosses, cited by Valérie Debuische, 'L'expression leibnizienne et ses modèles mathématiques', *Journal of the History of Philosophy* 51, no. 3 (2013): 435. Debuishce discusses Leibniz's alternate use of the terms 'concentration' or 'centre' to describe the monad.

27 'L'unité dans la multitude, voilà la principe architectonique du monde et de la philosophie, soit l'harmonie même, voilà, en termes inversés, le schéma élémentaire de la perception', Serres, *Le Système de Leibniz*, 96.

28 See Leibniz, *Monadology*, 56, 248.

29 Leibniz, Letter to Arnaud (1687) (G. ii. 112), quoted in *The Monadology and Other Philosophical Writings*, 136.

30 See Leibniz, 'What is an Idea' in Leibniz, Gottfried Wilhelm. *Philosophical Papers and Letters.* Edited by Leroy E. Loemker. 2. ed. Synthese Historical Library 2. Dordrecht: Kluwer, 1989, 207, and *Sur le principe de raison*, C11 in Leibniz, Gottfried Wilhelm. *Recherches générales* sur l'analyse des notions et des vérites: 24 thèses métaphysiques et autres textes logiques et métaphysiques. Translated by Jean-Baptiste Rauzy. 1. éd. Épiméthée. Paris: Presses Univ. de France, 1998, quoted in Debuische, 'l'expression leibnizienne', 413–414.

31 Debuiche describes the scholarly debate over the precise meaning of Leibniz's concept of expression as revolving around two main points. First, the precise nature of the relation of expression, i.e. whether it is a functional relation, or a relation of correlation, as Swoyer defines it, a 'structure preserving mapping'; and second, which elements from the expresser are preserved in the expressed. In her own interpretation, the relation of expression consists of an isomorphism in the strong sense: two things that possess the same form, without being exactly the same. Debuiche, 'L'expression leibnizienne', 415, 437. Chris Swoyer, 'Leibnizian Expression', *Journal of the History of Philosophy* 33 (1995): 82, quoted in Debuiche, ibid.

32 See Gottfried Wilhelm Leibniz, 'Correspondence with Arnaud, 1686–87', letter from 9 October 1687, in *Philosophical Papers and Letters*, 339.

33 Leibniz, 'What is an Idea?', ibid.

34 Leibniz, 'Principles of Nature and Grace', 409.

35 Leibniz, *Monadology*, 14, 224 modified.

36 See 'The Monadology', 14, *Philosophical Essays*, 214. This distinction enabled Leibniz to develop a complex theory of unconscious which we will discuss in Chapter Two.

37 See Leibniz, *New Essays*, Book of Ideas, 231.

38 Leibniz, 'What is an Idea?', *Philosophical Papers and Letters*, 207.

39 Leibniz, 'What is an Idea?', ibid, 207–8.

40 Ibid, 208.

41 Leibniz, *Monadology*, 56, in *The Monadology and Other Philosophical Writings*, 248.

42 See Pierre Levy, *Becoming Virtual: Reality in the Digital Age* (New York: Plenum Trade, 1998), 33.

43 Leibniz, *Monadology*, 7, 219.

44 As Leibniz writes, 'thus there is a perfect harmony, between the perceptions of the monad and the motions of the bodies, preestablished from the first between the system of efficient causes and that of final causes. And in this consists the agreement and the physical union of soul and body, without the one being able

to change the laws of the other.' 'Principles of Nature and Grace', 3, *Philosophical Essays*, 208.

45 See Benjamin, *Origin of the German Trauerspiel*, 26–7.

46 Leibniz, Lettre à Arnauld (1687) (G. ii. 95), quoted in *Monadology and Other Philosophical Writings*, 47.

47 Leibniz, *Monadology* 51, *Monadology and Other Philosophical Writings*, 246.

48 Leibniz, 'Principles of Nature and Grace', 10, *Philosophical Essays*, 210.

49 As Leibniz writes in 'Principles of Nature and Grace', 18, happiness consists in progression towards pleasures and perfections, rather than in the pleasures themselves. ibid, 213.

50 See the texts assmebled in Gottfried Wilhelm Leibniz, *De l'horizon de la doctrine humaine: 1693*, ed. Michel Fichant, Bibliothèque des textes philosophiques (Paris: Librairie philosophique), 1991.

51 See Leibniz, 'Monadology', 88–9, *Monadology and Other Philosophical Writings*, 269.

52 'Every notion of time and space we have is grounded in this accord', as Leibniz writes, quoted in Heinrich Schepers, 'Space and Time', in *The Oxford Handbook of Leibniz*, ed. Maria R. Antognazza (Oxford: Oxford University Press, 2018), 412.

53 'I hold space to be something merely relative, as time is: I hold it to be an order of co-existences, as time is an order of succession. For space denotes, in terms of possibility, an order of things which exist at the same time, considered as existing together; without inquiring into their particular manner of existing. And when many things are seen together one perceives that order of things among themselves.' Letter to Clark (Clarke's tr.) (E. 752 a; G. vii. 363): quoted in Latta, *The Monadology and Other Philosophical Writings*, 102.

54 See Schepers, 'Space and Time', 413.

55 See John T. Roberts, 'Leibniz on Force and Absolute Motion', *Philosophy of Science* 70, no. 3 (July 2003): 553.

56 Schepers argues that the crucial point of Leibniz's theory of time starts in 1676; however, the definition of time as an 'order' does not appear until 1685–6. See Schepers, 'Space and Time', 416.

57 Ibid, Thesis XVII, 396.

58 Walter Benjamin, *Über den Begriff der Geschichte*, ed. Gérard Raulet 1. Aufl, Werke Und Nachlaß, Bd. 19 (Berlin: Suhrkamp, 2010), 28. Recently, Dipesh Chakrabarty has focused on the co-existence of various scales of history within Benjamin's concept of 'now time' as a foundation for his concept of 'deep time' that captures the 'recognition of the otherness of the planet and its very large scale spatial and temporal processes of which we have unintentionally, become a part'. *The Human Condition in the Anthropocene,* The Tanner Lectures in Human Values, Yale University, February 18–19, 2015, 181.

59 Deleuze, *Expressionism in Philosophy*.
60 Ibid, 19.
61 Deleuze, *Expressionism in Philosophy*, 175.
62 Ibid, 17.
63 Ibid, 5.
64 Ibid, 62, 358.
65 Ibid, 322, modified.
66 Ibid.
67 'For, in the end, what is man in nature? A nothing compared to the infinite, an everything compared to the nothing, a midpoint between nothing and everything, infinitely removed from understanding the extremes: the end of things and their principles are hopelessly hidden from him in an impenetrable secret. equally incapable of seeing the nothingness from which he derives and the infinite in which he is engulfed.' Blaise Pascal, *Pensées*, trans. Roger Ariew (Indianapolis, IN: Hackett Pub. Co, 2005), 59.
68 Leibniz, 'The Principles of Philosophy, or, the Monadology', *Philosophical Writings*, 216/*La Monadologie*: 14. In some translations, '*gros de l'avenir*' is translated as 'big with the future', e.g. Leibniz and Strickland, *Leibniz's Monadology*, 76. Both translations are correct but it seems the intended meaning was closer to 'pregnant'.
69 Leibniz, 'First Truths', *Philosophical Essays*, 32.
70 Ibid, 16. See further discussion of implication and explication on p. 113.
71 Benjamin, 'On Language as such and the Language of Man', *SW1*, 69.
72 Ibid, 109.
73 Ibid, 80–1.
74 Latta explains this in the following: 'the work of the calculus is to determine the relations between unknown quantities or magnitudes, not by considering them merely as fixed wholes and directly finding equations between them, but indirectly, by treating the quantities as variables or as growing, and in the first place finding equations between their elements or their differences.' In the background for the development of differential calculus lies the shift from synthetic to analytic geometry, starting with Kepler's introduction of infinity into geometry. 'Statement of Leibniz's Philosophy', *Monadology and Other Philosophical Writings*, 81.
75 Deleuze, *Difference and Repetition*, 29/*Différence et répétition*, Bibliothèque de philosophie contemporaine. Paris: Presses Universitaires de France, 1981, 4.
76 Ibid, 261/341.
77 Ibid, 133/174.
78 Deleuze, *Difference and Repetition*, 171/*Différence et répétition* 222.
79 Leibniz, 'Justification of the Infinitesimal Calculus by that of Ordinary Algebra' (1701), in *Philosophical Papers and Letters*, 545–6.
80 Deleuze, *Difference and Repetition*, 51.

81 However, as Alex Tissandier writes, 'while he criticises Leibniz for subjecting this world to conditions of convergence, the very notions of convergence and divergence become central to Deleuze's own account'. See *Affirming Divergence*, 6.

82 Ibid, 52.

83 Daniel W. Smith, 'Mathematics and the Theory of Multiplicities: Deleuze and Badiou Revisited', in *Essays on Deleuze*, ed. Daniel W. Smith (Edinburgh: Edinburgh University Press, 2012), 293.

84 Ibid, 294. In the years since, some of the seventeenth-century principles of calculus have been redeemed. See John L. Bell, 'Continuity and Infinitesimals', in *The Stanford Encyclopedia of Philosophy* (Spring 2022 Edition), ed. Edward N. Zalta, First published Wed Jul 27, 2005; substantive revision Wed Mar 16, 2022, https://plato.stanford.edu/archives/spr2022/entries/continuity.

85 Smith compares between Deleuze's reliance on Leibniz's differential calculus and Alain Badiou's set-theory informed concept of multiplicity, ibid.

86 Beatrice Hanssen, 'Philosophy at Its Origin: Walter Benjamin's Prologue to the *Ursprung des deutschen Trauerspiels*', *MLN* 110, no. 4 (September 1995): 820.

87 See Walter Benjamin, 'Zur Sprachphilosophie und Erkenntniskritik', in *GS* 6, 9–53; Benjamin, 'On the Program for the Coming Philosophy', in *SW*1, 100–10.

88 Benjamin, *Origin of the German Trauerspiel*, 1/GS1, 207.

89 See Benjamin, *Arcades Project* [N2a,3], 462.

90 Benjamin, *Origin of the German Trauerspiel*, 42.

91 For example in the essay on Goethe: 'Goethes Wahlverwandtschaften', in *GS*1, 123–202/'Goethe's Elective Affinities', in *SW*1, 297–360.

92 See Benjamin, 'Doctrine of the Similar', in *SW*2, 696–7.

93 Ibid, trans. modified: *Lehre* may be translated as either 'doctrine' or 'teaching'. I find the latter to be more fitting in this context.

94 Benjamin, *Origin of the German Trauerspiel*, 3. *Darstellung* may be translated as either 'presentation' or 'representation', unlike the German *Repräsentation*. In a note the translators explain that by *Glasfluss* Benjamin refers to 'lead glass colored with metallic oxides. It has been used since antiquity to imitate precious stones'. Ibid/*GS* 1, 208.

95 Julia Ng, 'Acts of Time: Cohen and Benjamin on Mathematics and History', *Paradigmi. Rivista di critica filosofica* 1 (2017): 41–60.

96 Benjamin, *Origin of the German Trauerspiel*, 13.

97 Ibid, 23–4. As Benjamin writes of Croce's *The Essence of Aesthetic*: 'With this obscure sentence the author touches – unfortunately, all too hurriedly – upon the core of the theory of ideas. He is kept from full awareness of it by a psychologism that dissolves his definition of art as 'expression' into that of art as "intuition"'. See Croce Benedetto, *Essence of Aesthetic* (London: Heinemann, 1921).

98 Ibid, 27.

99 Ibid, 20.

100 Ibid, 27.

101 As Benjamin writes, 'not infrequently, ignorance of this discontinuous finitude has crippled energetic attempts to renew the theory of ideas, including, in the end, that of the early Romantics. In their speculations, truth assumed, instead of its linguistic character, the character of a reflecting consciousness'. Ibid, 15.

102 Ibid, 26.

103 Ibid, 13.

104 As Benjamin writes, 'even then, the danger of allowing oneself to plunge from the heights of knowledge to the monstrous depths of the Baroque temper is not to be disdained. Again and again one encounters, in the improvised attempts to realize the meaning of this epoch, that characteristic feeling of vertigo induced by the spectacle of the epoch's spirituality, as it revolves in contradictions.' Ibid, 39.

105 Ibid, 14.

106 This makes manifest the extent to which Deleuze's reading of Leibniz is informed by Belaval and Serres, both of whom focused on Leibniz's mathematical models and the manner in which they ground his metaphysics. See Serres, *Le Système de Leibniz*; Yvon Belaval, *Leibniz: initiation à sa philosophie*, 7. éd, Bibliothèque d'histoire de la philosophie (Paris: Vrin, 1993), *Études leibniziennes: de Leibniz à Hegel*, Bibliothèque des idées (Paris: Gallimard, 1976) and *Leibniz critique de Descartes*.

107 Deleuze, *Difference and Repetition*, 208.

108 Deleuze, *Difference and Repetition*, 205 fn./*Différence et répétition* fn. 20, 322.

109 See Leibniz, *New Essays*, book iv, 1, 2. Serres refers to this element of Leibniz's thought as his 'method of in-determination'. See Serres, *Le Système de Leibniz*, 221.

110 Ibid.

111 Ibid.

112 Leibniz, 'What Is an Idea?', *Philosophical Papers and Letters*, 207.

113 As Benjamin wrote in a curriculum vitae from 1928, his goal was to 'bring about a process of integration in scholarship – one that will increasingly dismantle the rigid partitions between disciplines that typified the concept of the sciences in the nineteenth century – and promote this through an analysis of the work of art', cited in Tagliacozzo, *Experience and Infinite Task*, 100. Deleuze challenges disciplinary boundaries in the *Cinema* books for example, where he compares film directors to philosophers. On the other hand, in *What Is Philosophy?* He separates between art, philosophy and science through the distinction between percept, affect and concept. See Deleuze and Guattari, *What Is Philosophy?*, 'Percept, Affect and Concept', 163–200.

114 Deleuze and Guattari, *What Is Philosophy?*, 17.

115 Benjamin, *Origin of the German Trauerspiel*, 56 modified/*GS*, 237.

116 Deleuze and Guattari, *What Is Philosophy?*, 23 modified/Gilles Deleuze and Félix Guattari, *Qu'est-ce que la philosophie?* (Paris: Éd. de Minuit, 1991), 29.

117 Benjamin, *Origin of the German Trauerspiel*, 2.

Chapter 2

1 Deleuze, *Difference and Repetition*, 166 (moderated)/*Différence et repétition*, 215.

2 Ibid, 130.

3 Ibid, 134.

4 Ibid, 136.

5 See Leibniz, *La naissance du calcul différentiel*, ed. Marc Parmentier and Michel Serres, Mathesis (Paris: J. Vrin, 1989) for a comprehensive collection of Leibniz's papers on calculus. For an account of how Leibniz developed his calculus, see Richard C. Brown, 'The Creation of Calculus', in *The Tangled Origins of the Leibnizian Calculus: A Case Study of a Mathematical Revolution* (Singapore; Hackensack, NJ: World Scientific, 2012).

6 Deleuze, *Difference and Repetition*, 161.

7 Deleuze's analysis of the differential in *Difference and Repetition* is not at all 'decoupled from the Leibnizian system in which Deleuze finds it', as Posteraro writes in 'Not Just Do as I Do: Knowledge and Learning in the Image of Thought', *Deleuze Studies* 9, no. 4 (2015): 459. To the contrary, it is grounded in his reading of Leibniz. Deleuze's emphasis on Leibniz's differential manifests the extent to which his reading of Leibniz is informed by Belaval and Serres. See Serres, *Le Système de Leibniz*; Belaval, *Leibniz*, *Études leibniziennes* and *Leibniz critique de Descartes*. As David Holdsworth mentions, Deleuze's 'remarks about learning and the master-pupil relation are often embedded in a context where he is discussing mathematics'. See 'Philosophical Problematisation and Mathematical Solution: Learning Science with Gilles Deleuze', *Deleuze and Education*, Deleuze Connections, ed. Inna R. Semetsky and Diana Masny (Edinburgh: Edinburgh University Press, 2013).

8 Deleuze, *Difference and Repetition*, 162–163/208.

9 Deleuze, ibid, 322 fn. 20/Différence et répétition 205 fn., 212.

10 See Leibniz, *New Essays*, book iv, 1, 2; 355–7. Serres refers to this element of Leibniz's thought as his 'method of in-determination'. See Serres, *Le Système de Leibniz*, 221.

11 Leibniz, ibid.

12 Deleuze, ibid, 163.

13 Leibniz, 'What is an Idea', *Philosophical Papers and Letters*, 207.

14 Deleuze, ibid, 162.

15 Ibid, 208–9.

16 See Postero, 'Not Just Do as I Do', 465.

17 At the same time, Deleuze criticizes Leibniz's concept of the virtual for its hesitation between the virtual and the possible; see Deleuze, ibid, 213.

18 Deleuze, ibid, 164.

19 For example, in Kant's first antinomy, where Kant himself refers to philosophers of the 'Leibnizian school' and their conception of the infinite. See Kant, *Critique of Pure Reason*, 473.

20 Deleuze, *Cours Vincennes*, cours du 06/05/1980.

21 See David Blumenfeld, 'Leibniz on Contingency and Infinite Analysis', *Philosophy and Phenomenological Research* XLV, no. 4 (June 1985): 484.

22 See Blumenfeld, ibid, 488–9.

23 Leibniz, 'On Freedom', in *Philosophical Papers and Letters*, 265–6.

24 Hermann Cohen, *Das Prinzip der Infinitesimal-Methode und seine Geschichte, ein Kapitel zur Grundlegung der Erkenntniskritik* (Frankfurt am Main: Suhrkamp, 1968), 57f., quoted in Alberto Poma, *The Critical Philosophy of Hermann Cohen*, trans. John Denton (Albany: State University of New York Press, 1997), 39.

25 Cohen, quoted in Poma, *Critical Philosophy of Hermann Cohen*, 195.

26 Ibid, 197.

27 As Cohen writes, 'God's love of man is the arche-type *(Urbild)* for man's moral task to love his fellowman, and, inasmuch as it is a task, love of fellowman is an infinite movement toward God as the supreme end', *RV*, 170; Eng. trans. Poma, *Critical Philosophy of Hermann Cohen*, 147.

28 Deleuze, *Cours Vincennes*, 20/05/1980.

29 Ibid.

30 Ibid.

31 Benjamin, 'On the Program of the Coming Philosophy', in *SW* 1, 108/*GS*2, 168.

32 Between 1913 and 1915 he studied at the University of Berlin under Ernst Cassirer, Benno Erdmann and Georg Simmel, and at the University of Munich, where he took lessons on Kant's *Critique of Judgment* with the phenomenologist Moritz Geiger. He also read and discussed with Scholem Kant's *Critique of Pure Reason*. See Tagliacozzo, *Experience and Infinite Task*, 12.

33 See Gershom Scholem, *Walter Benjamin: The Story of a Friendship* (New York: New York Review Books, 2012), 59.

34 See Tagliacozzo, *Experience and Infinite Task*, esp. 14; Ng, 'Acts of Time', 41–60.

35 As Benjamin writes, 'philosophy always inquires about knowledge, in relation to which the question of knowledge of its existence [*Dasein*] is only a modification … of the question of knowledge in general', 'On the Program of the Coming Philosophy', *SW*1, 109.

36 Benjamin, *SW*1, 100.

37 Ibid, 100–101.

38 Ibid, 101.

39 Ibid, 107.

40 Ibid.

41 Ibid. Reformulating Kant's concept of experience was, for Benjamin, a lifelong task, as Howard Caygill has shown in *Walter Benjamin: The Colour of Experience* (London; New York: Routledge, 1998), xii–xiii.

42 See Benjamin, 'Duns Scotus', *GS*6, 22–3.

43 Ibid.

44 Benjamin, *SW*1,108.

45 Hermann Cohen *Logik der reinen Erkenntnis*, System Der Philosophie, 1, 1. Aufl. (Berlin: Bruno Cassirer, 1902). See Tagliacozzo, *Experience and Infinite Task*, 29.

46 Leibniz's infinite analysis is the basis for Kant and Cohen's conceptions.

47 *GS*6, 51/translated in Tagliacozzo, *Experience and Infinite Task*, 32, modified.

48 Ibid, 52.

49 Ibid, 53.

50 Ibid.

51 Ibid.

52 Benjamin, *SW*1, 109–10.

53 'Continuity became the higher concept, from which infinity derives [...] For the being of infinity is now grounded in the thought of continuity. Continuity, in as much as it is idea, and in as much as it is law, is now a principle both of reason and of nature', as Cohen wrote in *Das Prinzip der Infinitesimal-Methode und seine Geschichte*, 57f., quoted in Poma, *The Critical Philosophy of Hermann Cohen*, 39.

54 The totalization of experience resulting in teachings or their equivalent, philosophy, is formulated in another fragment, 'On Perception', as the following: 'Philosophy is absolute experience, deduced in the systematic-symbolic context as language.' See Benjamin, *SW*1, 96.

55 Walter Benjamin, Gershom Scholem, and Theodor W. Adorno, *The Correspondence of Walter Benjamin, 1910–1940* (Chicago: University of Chicago Press, 1994), 98.

56 Scholem, *A Story of a Friendship*, 139.

57 *GS*6, 43–45/'Analogy and Relationship', in *SW*1, 207–9.

58 Benjamin, 'Goethe's Elective Affinities', *SW*1, 340.

59 Benjamin, *SW*1, 208/*GS*6, 44.

60 Scholem, quoted in *GS* 6, 658.

61 In notes for his *Habilitationsschrift*, under the title 'Theorems about symbolism', Benjamin lists the following:
1) Nothing objective as such corresponds with God
2) Nothing objective and nothing symbolic reaches God
3) Certain objects fulfil themselves only in an associated objective intention [*objektiven Intention*] and then point to God. These are objects of the highest order. *SW1*, 269, *GS* 6, 21–2.

62 Benjamin, 'On Perception', *SW* 1, 96. Benjamin explicitly appeals to a speculative conception of experience therein.

63 Benjamin, 'Theory of Knowledge', *SW* 1, 276 modified, Benjamin's italics/*GS* 6, 45.

64 Ibid.

65 'It is not in poetic analogies that Goethe discovered the symbols in which nature is recognizable, but in his prophetic insights. The *Urphänomen* is a systematic-symbolic concept', wrote Benjamin in a fragment from 1918. Benjamin, *GS* 6, 38.

66 See Helmut Holzhey, *Natorp und Cohen* (Basel: Schwabe, 1986), 184f, and Poma, *The Critical Philosophy of Hermann Cohen*, 293n.

67 Cohen, *Logik der reinen Erkenntnis* 4. Auflage, 3. Nachdruckauflage; Nachdruck der 2. verbesserten Auflage Berlin von 1914. Werke, Hermann Cohen. Bd. 6, ed. Helmut Holzhey (Hildesheim: Olms, 2005), 23, my translation.

68 Cohen, *Logik der reinen Erkenntnis*, 35, quoted in Andrea Poma, *The Critical Philosophy of Hermann Cohen* (Albany: State University of New York Press, 1997), 92.

69 See Marco Giovanelli, *Reality and Negation – Kant's Principle of Anticipations of Perception: An Investigation of Its Impact on the Post-Kantian Debate*, Studies in German Idealism, vol. 11 (Dordrecht Heidelberg London New York: Springer, 2011), 204.

70 Poma, *The Critical Philosophy of Hermann Cohen*, 95.

71 Ibid.

72 Cohen, *Logik der reinen Erkenntnis*, 36.

73 Ibid, 33.

74 Cohen, *Logik der reinen Erkenntnis*, quoted in Thomas Mormann and Mikhail Katz, 'Infinitesimals as an Issue of Neo-Kantian Philosophy of Science', *HOPOS: The Journal of the International Society for the History of Philosophy of Science* 3, no. 2 (Fall 2013): 244. See Cohen, *Das Prinzip der Infinitesimal-Methode und seine Geschichte, ein Kapitel zur Grundlegung der Erkenntniskritik*, 59, *Logik der reinen Erkenntnis*, 35.

75 See Cohen, *Das Prinzip der Infinitesimal-Methode*, 55.

76 Ibid.

77 Cohen, *Logik der reinen Erkenntnis*, 36, my translation.

78 Ibid, 41.

79 Benjamin, *Origin of the German Trauerspiel*, 24.

80 Ibid, 25.

81 Benjamin, *GS*1, 935, translation quoted in Hanssen, 'Philosophy at Its Origin', 824, modified.

82 Benjamin, *Origin of the German Trauerspiel*, 25.

83 Leibniz, 'The Principles of Philosophy, or, the Monadology', *Philosophical Writings*, 216.

84 See Leibniz, 'The Monadology', 12: 'But besides the principle of change, there must be a particular series of changes [*Mais il faut aussi qu'outre le principe du changement il y ait un detail de ce qui change*] which constitutes, so to speak,

the specific nature and variety of simple substances'. *The Monadology and Other Philosophical Writings*, 223. Leibniz, *La Monadologie (1714)*, 12. Leibniz also writes that 'each individual substance always contains traces of what has ever happened to it and marks of what will ever happen to it'. 'Remarks on Arnauld's Letter about my Proposition that the Individual Notion of Each Person Includes Once and for All Everything that Will Ever Happen to Him' [May 1686], *Philosophical Essays*, 71.

85 Deleuze, *The Fold*, 86/*Le Pli*, 235.

86 Deleuze, *Difference and Repetition*, 165.

87 Leibniz, 'Preface', *New Essays*, 55.

88 Deleuze, *Difference and Repetition*, 88/Différence et répétition, 239.

89 Deleuze, *The Fold*, 88 modified/*Le Pli*, 117.

90 See Leibniz, 'Justification of Infinitesimal Calculus by that of Ordinary Algebra (1701)', in *Philosophical Papers and Letters*, 542–3.

91 See Deleuze, *Difference and Repetition*, 173–4; *The Fold*, 89–90. See Salomon Maimon, *Essay on Transcendental Philosophy* (London and New York: Continuum, 2010).

92 Daniela Voss, 'Maimon and Deleuze: The Viewpoint of Internal Genesis and the Concept of Differentials', *Parhesia* 11 (2011): 63. See also Daniel W. Smith, 'Genesis and Difference: Deleuze, Maimon, and the Post-Kantian Reading of Leibniz', in *Deleuze and the Fold; A Critical Reader*, ed. Niamh McDonnell and Sjoerd van Tuinen (London: Palgrave Macmillan, 2009), 132–54, Graham Jones, 'Solomon Maimon', in *Deleuze's Philosophical Lineage*, ed. Graham Jones and Jon Roffe (Edinburgh: Edinburgh University Press, 2009), 104–29.

93 Deleuze, *Difference and Repetition*, 173. Maimon, *Essay on Transcendental Philosophy*, 38.

94 Maimon, ibid, 63.

95 See Samuel Hugo Bergman, 'Salomon Maimon', in *A History of Philosophy, Systems of Post-Kantian Philosophy* [תולדות הפילוסופיה החדשה : שיטות בפילוסופיה שלאחר קאנט] (Bialik: Jerusalem, 1979), 27–14.

96 In a footnote, Maimon justifies his use of mathematical concepts in order to explain philosophical idea by saying that 'these concepts in fact belong to philosophy and were taken from there into mathematics; as well as that the great Leibniz came upon the discovery of differential calculus through his system of the Monadology'. Maimon, Ibid, 19.

97 Ibid, 21.

98 Ibid, 19.

99 Ibid, 20.

100 Deleuze, *The Fold*, 89 (modified)/*Le Pli*, 119.

101 Deleuze, *Difference and Repetition*, 222.

102 Kant, *Critique of Pure Reason*, B207.

103 Ibid, A170/B211.

104 Maimon develops this idea of 'flowing' within internal cognition rather than applying the concept to real objects: 'the understanding cannot think of the object as having already arisen[*entstanden*] but only as arising [*entstehend*], i.e as flowing [*fliessend*]'. Ibid, 22.

105 Deleuze, *Difference and Repetition*, 281.

106 Immanuel Kant, 'On the Basis of the Difference of Regions in Space' (1768), in *Gesammelte Schriften*, ed. Akademie der Wissenschaften (Berlin: Reimer, de Gruyter, 1900), vol. 2, 378.

107 He refers to the distinction between *innere Verschiedenheit* and *innerlich* in Kant's 'Prolegomena', 13 in Immanuel Kant, Gary C. Hatfield, and Immanuel Kant, *Prolegomena to Any Future Metaphysics That Will Be Able to Come Forward as Science: With Selections from the Critique of Pure Reason*, Rev. ed., Cambridge Texts in the History of Philosophy (Cambridge, UK; New York: Cambridge University Press, 2004).

108 Deleuze, *Difference and Repetition*, 231.

109 Ibid, 26.

110 Ibid, 165/214.

111 See also Deleuze, *Cours Vincennes*, cours du 29/4/1980: https://www.webdeleuze.com/textes/54.

112 Deleuze, *Difference and Repetition*, 148 (modified)/*Différence et répéptition*, 192.

113 Ibid, 214.

114 Ibid, 326/253, modified.

115 Ibid, 246/317. See Stefano Di Bella, *The Science of the Individual: Leibniz's Ontology of Individual Substance* (Dordrecht: Springer Netherlands, 2005).

116 Ibid, 247/318.

117 *Biographie croisée*, 534. Dosse cites Christine Buci-Glucksmann, 'Les cristaux de l'art: une esthétique du virtuel', *Rue Descartes* 20 (May 1998): 95–111.

118 Deleuze, *Difference and Repetition*, 247/*Différence et repetition*, 318. See also Deleuze's discussion of implication and explication as two ways in which meaning is contained in the sign in Gilles Deleuze, *Proust and Signs* trans. Richard Howard (London, New York: Continuum, 2008).

119 Ibid, 335.

120 Leibniz refers to *Meno* in *Discourse on Metaphysics*, one of the few texts by Leibniz that Benjamin cites. In Leibniz's interpretation, the soul recollects data not from previous lives, but from periods in which it is inactive, or does not think: 'our soul always has in it the quality of representing to itself any nature or form whatsoever, when the occasion to think of it presents itself. And I believe that this quality of our soul, in so far as it expresses some nature, or essence, is properly the idea

of the thing, which is in us and which is always in us, whether we think of it or not.' Leibniz, 'Discourse on Metaphysics', 26, Leibniz, *Philosophical Essays*, 58.

121 Deleuze, *Difference and Repetition*, 114–24, quote on 114.

122 Benjamin, *Origin of the German Trauerspiel*, 13/*GS* 1, 217.

123 Ibid.

124 See 'On Language as Such and the Language of Man', *SW1*, 69.

125 Ibid, 72.

126 See for example Deleuze, *Logic of Sense*, 253.

127 Deleuze, *Difference and Repetition*, 167.

128 See Leibniz, 'Discourse on Metaphysics', 26, *Philosophical Essays*, 58.

129 Ibid.

Chapter 3

1 Benjamin, *Arcades Project*, [N11,4], 476 modified/ *GS*5, 596.

2 See Benjamin, letter to Gershom Scholem from 20 January 1930. *The Correspondences*, 359. The 'Convolutes' are thematic sections of the *Arcades Project* that revolve around cultural and social phenomena of nineteenth-century Paris. The term *Konvolut*, which means an assemblage or bundle of texts, was apparently chosen by Adorno, who was the first to review the manuscript which was hidden during the Second World War in the Biblioteque National in Paris by George Bataille. See Howard Eiland and Kevin McLaughlin, 'Translators Forward', *The Arcades Project*, xiv.

3 Ibid, [N,10a,3], 475. Benjamin's use of the terms 'historical materialism' and 'dialectics' is specific and idiosyncratic. As Löwy's suggests 'taking its inspiration from Marxist and messianic sources, it uses nostalgia for the past as a revolutionary method for the critique of the present', 'Introduction: Romanticism, Messianism and Marxism in Walter Benjamin's Philosophy of History' in *Fire Alarm: Reading Walter Benjamin's 'On the Concept of History'* (London: Verso, 2016), 13. When Benjamin refers to the 'historical materialist', he sets a standard for the kind of historiography the historian should strive for, and at the same time refers to his own historical-cultural project of the *Arcades Project*. As in the Foreword to *Origin*, many of Benjamin's later references to 'history' concern the history of art. See, for example: 'for the dialectical historian concerned with works of art, these works integrate their fore-history as well as their after-history; and it is by virtue of their after history that their fore-history is recognizable as involved in a continuous process of change'. Walter Benjamin, 'Eduard Fuchs, Collector and Historian', in *SW*3, 261.

4 This division has a long history in the reception of Benjamin, as discussed in the Introduction.

5 As discussed, Leibniz's understanding of substance changed over time. In later work, he viewed action as derivative of substance, as Schepers discusses: 'matter and motion are not so much substances or things in themselves as the phenomena of the perceivers.' Heinrich Schepers, 'Monadic Perception', in *The Oxford Handbook of Leibniz*, ed. Maria Rosa Antognazza (Oxford: Oxford University Press, 2018), 381.

6 Leibniz also describes a passive force that moves in the reverse direction, as will be discussed.

7 Carl Schmitt, *Political Theology: Four Chapters on the Concept of Sovereignty*, University of Chicago Press ed (Chicago: University of Chicago Press, 2005).

8 'In keeping with Russian folk belief, he interpreted the resurrection less as a transfiguration than a disenchantment', as Benjamin writes of Leskov, *SW*3, 158. See also *Arcades Project*, [N1a,3], 459.

9 See Maria Rosa Antognazza, 'Leibniz as Historian', in *The Oxford Handbook of Leibniz* (Oxford; New York: Oxford University Press, 2018), 607.

10 See Lewis W. Spitz, 'The Significance of Leibniz for Historiography', *Journal of the History of Ideas* 13, no. 3 (June 1952): 333. Leibniz's historical writing includes for example *gentium diplomaticus* (1693) followed by *Der origin Germanorum* (1696) and *Origines Guelficae* (1748). For a complete account see Louis Davillé, *Leibniz Historien, Essai sur l'activite et la methode Historiques de Leibniz* (Paris: Librairie Felix Alcan, 1909).

11 Davillé quotes Leibniz's contemporary Fontenelle, who makes a similar argument; see 687 (Eloge, Dutens, t.1, xxvi). His interpretation is evidently influenced by Historicist readings of Leibniz. See Davillé, *Leibniz Historien*, 666–74. More recently, Maria Rosa Antognazza has painted a more nuanced picture, suggesting that 'he conceived history as both providentially guided and following an overall trajectory of progress despite apparent cyclical set-backs … This providential direction followed, according to Leibniz, neither a simple cyclical path nor a straight line. Rather, it was best represented by the movement of an ascending spiral, visually combining cyclical return with unending progress.' 'Leibniz as Historian', 602. Leibniz uses the spiral image in 'Demonstrationes de Universo immenso aeternoque; de Mundis et aevis, deque rerum longiquarum et futurarum statu, c. February–March 1701', Leibniz, *De l'horizon de la doctrine humaine*, 58–9.

In 'Leibniz und sein Geschichtsbegriff' Thomas Leinkauf suggests that Leibniz's concept of history also included a notion of *historia naturalis*, or natural history. *Studia Leibnitiana*, Bd. 51, H. 1 (2019): 12–41.

12 See Allison Coudert, *Leibniz and the Kabbalah*. International Archives of the History of Ideas 142. Dordrect; Boston: Kluwer Academic, 1995. Leibniz

develops his ideas on return in three principal texts, dating from rather late in his career: 'De L'horizon de la doctrine Humaine', written in 1693; 'Apokatastasis', and 'Apokatastasis Panton', written in 1715, which are collected and edited by Fichant in the edition *De l'horizon de la doctrine humaine: 1693*. For a discussion of Benjamin and Deleuze's concepts of eternal return in relation to Leibniz's apokatastasis, see Noa Levin, 'Specters of Eternal Return: Benjamin and Deleuze read Leibniz', *Filozofski vestnik*, special issue on 'world' (2021): 305–29.

13 Coudert argues that Leibniz's last texts on apokatastasis embrace Origen's optimistic version of the term as the inevitable restitution of the fallen world to its prelapsarian perfection. Ibid, 114–15.

14 Gottfried Wilhelm Leibniz, *Theodicy: Essays on the Goodness of God, the Freedom of Man, and the Origin of Evil*, translated by E. M. Huggard and Austin Farrer (New York: Cosimo, 2009).

15 Immanuel Kant, 'Idea for a Universal History from a Cosmopolitan Perspective', in *Towards Perpetual Peace and Other Writings on Politics, Peace and History*, ed. Pauline Kleingeld, trans. David L. Colclasure (New Haven: Yale University Press, 2006), 14–15.

16 See Reinhart Koselleck, *Futures Past: On the Semantics of Historical Time* (New York: Columbia University Press, 2004), 266–7.

17 See Spitz, 'The Significance of Leibniz', 235–6; Koselleck, *Futures Past*, 266–8.

18 See Koselleck, ibid, 33–42.

19 'At first sight it is certainly a strange, and apparently an extravagant, project, to propose a History of Man founded on any idea of the course which human affairs would take if adjusted to certain reasonable ends. On such a plan it may be thought that nothing better than a romance could be the result.' Kant, 'Idea for a Universal History', 9. See Koselleck, *Futures Past*, 34–5.

20 Coudert follows Loemeker, Garber and Wilson, who claim that Leibniz's philosophy changed significantly in his later years, from the 1690s on, and that in the *Monadology* specifically it is evident that Leibniz's thought had undergone a revolution manifest in his concept of the dynamic monad in contrast to the static substance of the *Discourse* and earlier writings. See Coudert, *Leibniz and the Kabbalah*.

21 Coudert suggests that this influence was mainly mediated by the Christian theologian Francis Mercury van Helmont, with whom Leibniz had a close relation during the late years of his career. See ibid.

22 Benjamin's interest in the Kabbalah was mediated by his friend the Kabbalah scholar Gerschom Scholem, since he did not read Hebrew and had not read the Kabbalah. Scholem describes their discussions of Molitor's presentation of the Kabbala: Franz Joseph Molitor, *Philosophie der Geschichte oder Über die Tradition* (Münster: Theissing, 1857). See Scholem, *A Story of a Friendship*, 105. According

to McBride, what made the strongest impression on Benjamin about the Kabbalah was the theory presented therein of divine language. See James McBride, 'Marooned in the Realm of the Profane: Walter Benjamin's Synthesis of Kabbalah and Communism', *Journal of the American Academy of Religion* 57, no. 2 (Summer, 1989): 241–66.

23 Gottfried Wilhelm Leibniz, 'On the Correction of Metaphysics and the Concept of Substance', in *Philosophical Papers and Letters*, 433.

24 See Leibniz, *Monadology*, 18, *Monadology and Other Philosophical Writings*, 229.

25 Gottfried Wilhelm Leibniz, 'Specimen Dynamicum', in *Philosophical Papers and Letters*, 435–52.

26 Ibid, 435.

27 Benjamin describes life as unfolding in several early texts, such as the essay on Translation: 'Diese Enfaltung ist als die eines eigentümlichen und hohen Lebens durch eine eigentümliche und hohen Zweckmäßigkeit bestimmt'. See *GS* 4, 11.

28 Ibid, 13, 419.

29 An 'intermediate between bare potency and the fully developed activity [*acte*] of the scholastics', as Latta describes it in *Monadolodgy and Other Philosophical Writings*, 229. The word 'virtual' originates from Medieval Latin *virtualis* and Latin *virtus*, both of which mean potency, efficacy and literally manliness or manhood, and from which originates the English 'virtue'. The meaning 'being something in essence or effect, though not actually or in fact' is from mid-fifteenth century, see https://www.etymonline.com/word/virtual.

30 Leibniz, 'Introduction', in *New Essays, Monadology and Other Philosophical Writings*, 367.

31 Latta highlights that the notion of virtual and unconscious perceptions was developed against the positions of Locke and Descartes; see *Monadology and Other Philosophical Writings*, 367.

32 Leibniz, 'Introduction', 367–8.

33 G.W. Leibniz, 'Discourse on Metaphysics', *Philosophical Essays*, 35–68.

34 Ibid, 8, 41.

35 Leibniz, 'Remarks on Arnauld's Letter', ibid, 71.

36 Ibid, 69–76.

37 See Leibniz, 'On Freedom', 263.

38 Deleuze, *The Fold*, 52 (modified)/*Le Pli*, 70.

39 Plato, 'Meno', in *Meno and Other Dialogues*, Oxford World's Classics, trans. Robin Waterfield (Oxford; New York: Oxford University Press, 2005), 82b–86a, 115–23.

40 Leibniz, 'Discourse on Metaphysics', *Philosophical papers and Letters*, 26, 320.

41 The link between Benjamin's and Leibniz's conceptions of the virtual has been noted, for example by Kevin McLaughlin in 'Virtual Paris: Benjamin's Arcades Project', Richter, *Benjamin's Ghosts*, 204–35, who suggests that the significance of the term 'virtual' to Benjamin is influenced by Leibniz's use of the term.

42 Benjamin, *Origin of the German Trauerspiel*, 26–7.
43 Ibid, 26.
44 Benjamin, 'The Task of the Translator', *SW*1, 253–63.
45 Ibid, 224.
46 Ibid, 225.
47 See Adorno, 'Idea of Natural History'. Susan Buck-Morss interprets Benjamin's 'natural history' as a 'montage of nature and history' in a critical take on Social Darwinism. 'Natural History: Fossil', 58–77, Susan Buck-Morss, *The Dialectics of Seeing: Walter Benjamin and the Arcades Project*, 8. print (Cambridge, MA: MIT Press, 1999), 64.
48 Benjamin, 'To the Planetarium', *One-Way Street*, 93–6.
49 Ibid. See Theodor W. Adorno and Max Horkheimer, *Dialectic of Enlightenment*, New ed. (London: Verso, 2014).
50 Ibid, 46, 36.
51 This redemption, or salvation of ideas, through which their virtual history is constituted, takes place immanently, in a manner that corresponds with the immanent salvation, the only kind possible within *Trauerspiel* plays, since, as Benjamin discusses, these no longer contain an eschatological dimension. Ibid, 81.
52 Benjamin, *Origin of the German Trauerspiel*, 14/*GS*1, 217–18.
53 Johannes Kepler et al., *The Harmony of the World*, ed. Eric J. Aiton, Memoirs of the American Philosophical Society 209 (Philadelphia, Pa: American Philosophical Society, 1997). Benjamin mentions Kepler and other Early Modern astronomers as those who brought about a rift between nature and humanity in 'To the Planetarium', *One-Way Street*, 94.
54 Benjamin, *Origin of the German Trauerspiel*, 27.
55 Ibid.
56 Benjamin, 'Letter to Florens Christian Rang', *SW* 1, 387–90/Walter Benjamin, Christoph Gödde, and Henri Lonitz, *Gesammelte Briefe. 1. Aufl.* (Frankfurt am Main: Suhrkamp, 1995), 390–4.
57 Benjamin, ibid, *SW* 1, 389/*GB*, 393. When writing the letter, Benjamin was in the process of having his dissertation examined for his *Habilitation*, later published as the *Trauerspiel* book.
58 Ibid.
59 See for example Nicholas Rescher, *The Philosophy of Leibniz* (New Jersey: Prentice-Hall, 1967), 58–65.
60 As Leibniz writes in a letter to Clarke: 'I hold space to be something merely relative, as time is: I hold it to be an order of co-existences, as time is an order of succession', 'Lettre a Clarke' (Clarke's trans.) (E. 752 a; G. vii. 363), quoted in Leibniz, *Monadology and Other Philosophical Writing*, 102.
61 Benjamin, *SW*1, 389/*GS*, 393.
62 Gottfried Wilhelm Leibniz, 'On the Conception of Metaphysics and the Concept of Substance', in *Philosophical Papers and Letters*, 433.

63 Wilhelm Hausenstein, *Vom Geist des Barock* (München: Piper, 1920), 42, quoted in Benjamin, *Origin of the German Trauerspiel*, 51 (modified)/*GS*, 247.

64 Ibid, 92.

65 See Michael Futch, *Leibniz's Metaphysics of Time and Space* (New York: Springer, 2014), 194.

66 For a discussion of architectural and mechanical artifice in Baroque set design, especially those produced by architect and set designer Joseph Furttenbach, and how these correspond to Benjamin's analysis, see Michael Hosman, 'Benjamin's Baroque', *Thresholds*, no. 28 (2005): 119–29.

67 Benjamin, *Origin of the German Trauerspiel*, 49–50/*GS* 1, 246.

68 *Ausnahmezustand* has been translated as either 'state of exception' or 'state of emergency'. Giorgio Agamben discusses the concept and its use in what he describes as a debate between Schmitt and Benjamin, and, informed by this discussion, puts forward his own theory of biopolitical power in 'State of Exception', *The Omnibus Homo Sacer*, Meridian, Crossing Aesthetics (Stanford, California: Stanford University Press, 2017). Agamben hypothesizes that the debate between the two authors began when Schmitt read Benjamin's 'Critique of Violence' and responded to it in *Political Theology*, and continued in Benjamin's letter to Schmitt and Schmitt's reference to Benjamin in his *Hamlet or Hecuba: The Intrusion of the Time into the Play* (Candor, NY: Telos Press Publishing, 2014). See ibid, 212.

69 Schmitt, *Political Theology*. For analyses of the multiple meanings of the first sentence, see Foreword by Tracy B. Strong, *Political Theology: Four Chapters on the Concept of Sovereignty* and Hasso Hofmann, 'Souverän ist, wer uber die Ausnahmezustand Entscheidet', *Der Staat* 44, no. 2 (2005): 171–86.

70 For more on Schmitt's 'decisionism' and its foundations in Schmitt's 'friend-enemy' distinction outlined in *The Concept of the political*, see Paul Hirst, 'Carl Schmitt's Decisionism', and Chantal Mouffe, ed., *The Challenge of Carl Schmitt*, 1. publ. Phronesis (London: Verso, 1999).

71 Schmitt, ibid, 36.

72 See Peter Schröder, 'Carl Schmitt's Appropriation of the Early Modern European Tradition of Political Thought on the State and Interstate Relations', *History of Political Thought* 33, no. 2 (2012): 359.

73 Thomas Hobbes, *Leviathan*, ed. Richard Tuck, Cambridge Texts in the History of Political Thought (Cambridge; New York: Cambridge University Press, 1996). There are of course many differences, and for Schmitt the important early modern thinkers are French jurist Jean Bodin and Hobbes, whose utilitarian understanding of human nature and the natural condition diverge from Leibniz, who admired but also criticized Hobbes. For a discussion of Leibniz's position towards Hobbes, see Fiorenza Manzo, 'How sincere was Leibniz's criticism of

Hobbes' political thought?', *The Leibniz Review*, 2020, https://doi.org/10.5840/leibniz2020302.

74 See Benjamin's dedication of *Origin of the German Trauerspiel* to Schmitt, quoted in Samuel Weber, 'Taking Exception to Decision: Walter Benjamin and Carl Schmitt', in *Benjamin's -abilities* (Cambridge, MA; London: Harvard University Press, 2010), 176.

75 Benjamin, *Origin of the German Trauerspiel*, 48.

76 Ibid, 50.

77 Ibid, 49.

78 According to Agamben, Benjamin reformulates, in 'On the Concept of History', a distinction between a 'real' and a 'fictitious' state of exception made by Schmitt, yet 'Benjamin once again reformulates the opposition in order to turn it back against Schmitt. Now that any possibility of a fictitious state of exception – in which exception and normal conditions are temporally and locally distinct – has collapsed, the state of exception "in which we live" is real and absolutely cannot be distinguished from the rule. Every fiction of a nexus between violence and law disappears here: there is nothing but a zone of anomie, in which a violence without any juridical form acts'. See Agamben, 'Homo Sacer', ibid, 217.

79 Weber, ibid, 179. There have been many other readings that try to make sense of the exchange between Benjamin and Schmitt, such as Susanne Heil, '*Gefährliche Beziehungen*': *Walter Benjamin und Carl Schmitt* (Stuttgart: Metzler, 1996) and Marc de Wilde, 'Meeting Opposites: The Political Theologies of Walter Benjamin and Carl Schmitt', *Philosophy & Rhetoric* 44, no. 4 (2011): 363–81.

80 Benjamin, *Origin of the German Trauerspiel*, 11.

81 Schmitt, *Political Theology*, 5.

82 Benjamin, *Selected writings: 1927–1934*, vol. 2, ed. Marcus Paul Bullock et al. (Cambridge, MA: Harvard University Press 1999), 704. See also Rebecca Gould 'Laws, Exceptions, Norms: Kierkegaard, Schmitt, and Benjamin on the Exception', *Telos: Critical Theory of the Contemporary* 162 (2013): 77–96.

83 As Benjamin writes, 'The prince, with whom rests the decision concerning the state of exception, shows that, as soon as the situation arises, a decision is nearly impossible for him.' *Origin of the German Trauerspiel*, 56.

84 The exception as a cesura in continuous expression also echoes the concept of 'expressionless' that Benjamin formulated in earlier texts. As he writes in 'Goethe's Elective Affinities': 'The "Occidental Junoian sobriety" … is only another name for that cesura, in which along with harmony, every expression simultaneously comes to a standstill, in order to give free reign to an expressionless power inside all artistic media'. *SW1*, 341.

85 Benjamin, *Arcades Project*, [Q, 21], 867/*GS5*, 1038.

86 Ibid, 476.

87 Deleuze, *The Fold*, 10.
88 Benjamin, ibid.
89 See Leibniz, *Monadology*, 12: 'besides the principle of change, there must be *a particular series of changes* [*un detail de ce qui change*] which constitutes, so to speak, the specific nature and variety of the simple substances', *Monadology and Other Philosophical Writings*, 223.
90 Ibid, [O, 81], 865 modified/*GS*5, 1034–5.
91 Walter Benjamin, 'Attested Auditor of Books', in *One-Way Street* (Cambridge, MA: Belknap Press of Harvard University Press, 2016), 41–41/'*Vereidigter Bücherrevisor*', 'Einbahnstrasse', 1. Aufl, *Werke Und Nachlass/Walter Benjamin*, Bd. 8 (Frankfurt am Main: Suhrkamp, 2009), 29–31.
92 Benjamin, 'Attested Auditor', *One-Way Street*, 41.
93 Mallarmé, reproduced in Meillassoux, *The Number and the Siren*, 'Appendix 1', IX.
94 Ibid, I, XI.
95 'On his desk there was also Mallarmé's *Un Coup de dés* in a special quarto edition whose graphic form was clearly in keeping with the title. The words in various type sizes rolled back and forth over the lines like dice … The whole thing was a most astonishing sight, and Benjamin told me that he did not understand the text either. In my uncomprehending soul there remained engraved only the visual image of a pre-Dadaistic product.' Scholem, *A Story of a Friendship*, 202.
96 See Jacques Scherer, *Le "Livre" de Mallarmé* (Paris: Gallimard, 1998).
97 Stéphane Mallarmé, 'Préface' (Faksimile-Edition de l'édition, 1914; Paris: Gallimard, 2014), 3.
98 Benjamin, ibid, 42.
99 Paul Valéry, '*Le Coup de dés*'. *Variété* (Paris: Gallimard, 1998).
100 Leibniz, 'Lettre à Arnauld' (1687), (G. ii. 95), quoted in *The Monadology and Other Philosophical Writings*, 47.
101 Stéphane Mallarmé, 'La music et les lettres', in *Œuvres Complètes*, ed. Bertrand Marchal (Paris: Gallimard, 2003), 68.
102 See *Origin of the German Trauerspiel*, where Benjamin uses the term *Schriftbild* in his interrogation of the relation of the allegory to forms that combine image and script such as the emblem, rebus and hieroglyph, and suggests that during the Baroque '(e)xternally and stylistically – in the drastic character of the typography as in the over burdened metaphor – the written presses toward image. No starker contrast to the art symbol, the plastic symbol, the image of organic totality is conceivable than this amorphous fragment as which allegorical image writing [*Schriftbild*] reveals itself'. Ibid, 185. For further discussion of the relationships between image and script in Benjamin's *Origin of the German Trauerspiel* see section 'Allegory and Symbol', Chapter Five.

103 Meillassoux, 140–141.

104 Benjamin, ibid.

105 Benjamin, ibid, 93/Walter Benjamin et al., 'Einbahnstrasse', 1. Aufl, *Werke Und Nachlass/Walter Benjamin*, Bd. 8 (Frankfurt am Main: Suhrkamp, 2009), 30. The index card system has been since considered as a precursor of the database. See Lev Manovich, 'Database as Symbolic Form', *Convergence: The International Journal of Research into New Media Technologies* 5, no. 2 (1999): 80–99.

106 These are available at the Walter Benjamin Archiv, Academie der Künste, Berlin and reproduced on their website as 'Farbsignets zur Übertragung *Passagen/Baudelaire*': https://www.adk.de/de/archiv/archivabteilungen/walter-benjamin-archiv/index.htm.

107 Ibid, 457.

108 He alludes to 'Leibniz's concept language' in a list compiled in 1917–20, Benjamin-Archiv MS 506 varia.

109 See Gottfried Wilhelm Leibniz, 'Preface to a Universal Characteristic', in *Philosophical Essays*, Gottfried Wilhelm Leibniz, trans. Roger Ariew, and Daniel Garber (Indianapolis: Hackett Publishing Co., 1989), 9.

110 Gilles Deleuze, *Nietzsche and Philosophy. European Perspectives* (New York: Columbia University Press, 2006), 33.

111 Mallarmé reproduced in Meillassoux, *The Number and the Siren* 'Appendix 1', X, XI.

112 Deleuze, *The Fold*, 30.

113 Ibid.

114 Ibid, 3.

115 Ibid, 67.

116 Ibid, 68.

117 Deleuze, Gilles, *Nietzsche and Philosophy*, 33.

118 Ibid, 26.

119 Ibid, 29.

120 Benjamin, *The Arcades Project* [N1a,3], 459.

121 Blanqui's doppelganger world of 'present eternalised' is described as the 'phantasmagoria of history itself'. Ibid, 25–6.

122 For a reading that does make this connection, see Desideri, 'Intermittency: The Differential of Time and the Integral of Space, the Intensive Spatiality of the Monad', 177–87.

123 Benjamin, *Arcades Project* [N1a,3], 459/*GS*, 573, modified.

124 Benjamin, *SW*3, 143.

125 Benjamin, ibid, 158.

126 As Benjamin writes, 'Only a redeemed mankind … is granted the fullness of its past – which is to say, only for a redeemed mankind the past has become citable in all its moments.' Thesis III, *SW*4, 390.

127 Löwy, *Fire Alarm*, 83–4. As he writes, 'The Jewish, messianic and cabbalistic equivalent of the Christian apokatastasis is, as Scholem argues in his article "Kabbala" in the Encyclopaedia Judaica (1932), tikkun'.

128 Ibid.

129 Giorgio Agamben, *The Coming Community*, trans. M. Hardt (Minnesota: University of Minnesota Press, 1993), 54. Scholem notes in amusement in a letter from July 1934, 'One question: Who is actually the source of all these stories? Does Ernst Bloch have them from you or you from him? The great rabbi with the profound dictum on the messianic kingdom who appears in Bloch as well is none other than I myself; what a way to achieve fame! It was one of my first ideas about the Kabbalah.' See *The Correspondences*, 446.

130 Ibid.

131 See Leibniz, *De l'horizon de la doctrine humaine*, 35–67.

132 See Leibniz, *Monadology*, 88: 'A result of this harmony is that things lead to grace by the very ways of nature, and that this globe, for instance, must be destroyed and renewed by natural means at the very time when the government of spirits requires it, for the punishment of some and the reward of others.' 'Apokatastasis Panton' (1715). 'Hence it follows that if we imagine that humanity lasts long enough in the state it is in now, past public histories must return.' Leibniz, trans. David Forman, [64] 2.

133 As noted by David Forman, 1–2: 'certainly there will always be differences of some sort between the old and the new, but they will be imperceptible, and precisely of the sort that could not be expressed in books.'

134 In this way Leibniz can align repetition with progression. His struggle between the two positions is understandable, since on the surface, supporting eternal return would contradict the theory of the world's constant striving towards perfection, a main tenet of Leibniz's philosophy. See Gottfried Wilhelm Leibniz, 'La Restitution', in *De l'Horizon de la Doctrine Humaine*, 73.

135 Plato, Timaeus (39c–d).

136 Paul Ricœur, *Memory, History, Forgetting*, trans Kathleen Blamey and David Pellauer (Chicago: University of Chicago Press, 2010), 11. For Benjamin, it is the name rather than the image that forms part of the 'realm of the similar'; see Benjamin, *Arcades Project*, [Q, 24], 868.

137 Benjamin, 'Paralipomena to "On a Concept of History"', *SW* 4, 403–4.

138 Deleuze, *Essays Critical and Clinical*, trans. Daniel W. Smith and Michael A. Greco (London and NY: Verso, 1998), 27.

139 See Auguste Blanqui, *Eternity by the Stars: An Astronomical Hypothesis*, trans. Frank Chouraqui, First Contra Mundum Press edition (New York: Contra Mundum Press, 2013). Friedrich Nietzsche, R. J. Hollingsworth, and Reginald J. Hollingdale, *Thus Spoke Zarathustra: A Book for Everyone and No One*, Repr. with new chronology, Penguin Classics (London: Penguin Books, 1961).

140 Benjamin, *Arcades Project*, 25–6.
141 Ibid, 'Q, Panoramas', [Q1,1], 527.
142 Ibid, 25.
143 For a detailed discussion of how the concept of 'phantasmagoria' functions in the *Arcades Project* see Margaret Cohen, 'Walter Benjamin's Phantasmagoria', *New German Critique* 48 (Autumn 1989): 87–107.
144 Benjamin, *Arcades Project*, [J62a, 2], 341.
145 Ibid, 26.
146 Ibid, [J67,4], 348.
147 Ibid, 26.
148 Ibid, [J61a, 3], 339.
149 Ibid, [N9,5], 473.
150 By calling Nietzsche's Return a 'Copernican Revolution' Deleuze is of course alluding to Kant's self-proclaimed 'Copernican Revolution' in *Critique of Pure Reason*, 'Preface' Bxvi–xviii, 110–11.
151 Deleuze, *Difference and Repetition*, 40.
152 Ibid, 41.
153 Ibid, 42.
154 Ibid, 41.
155 Deleuze, *Nietzsche and Philosophy*, 68.
156 Ibid, 243.
157 Ibid, 41.
158 Ibid, 243.
159 For a reading of Nietzsche that elaborates his debt to Leibniz, see Günter Abel, *Nietzsche: Die Dynamik der Willen zur Macht und die Ewige Wiederkehr* (Berlin: De Gruyter, 1984).
160 Deleuze, ibid, 232, 236.

Chapter 4

1 See George Gentry, 'The Subject in Whitehead's Philosophy', *Philosophy of Science* 11, no. 4 (October 1944): 222–6.
2 For recent publications that examine perspectivism in the context of analytic philosophy and philosophy of science, see Michela Massimi, *Perspectival Realism*, Oxford Studies in Philos Science Series (New York: Oxford University Press, 2022) and *Understanding Perspectivism*, ed. Michela Massimi and Casey D. McCoy (London: Routledge, 2019).
3 See Debuiche, *L'expression leibnizienne*, 415.
4 See Belaval, *Études leibniziennes*, 144.

5 As formulated by Serres, 'La vision demeurant model de la connaissance, chose traditionnelle de Platon est Descartes, la théorie de la vision perspective, en tant qu'elle definit en rigeur la notion d'expression, est un genre dont la théorie de la perception et celle de la connaissance sont des espèces' Serres, *Le système de Leibniz*, 158.

6 For a detailed account of Leibniz's longstanding research in the field, see Jeffrey K. McDonough, 'Optics', in *The Oxford Handbook of Leibniz*, ed. Maria Rosa Antognazza (Oxford: Oxford University Press, 2018), 425–37. A thriving science in the seventeenth century, optics was the focus of several research projects undertaken by Leibniz. As McDonough shows, Leibniz made innovative attempts at deriving the laws of reflection and refraction throughout his career. McDonough traces Leibniz's optical research from the early *Leges Reflexionis et Refractionis Demonstratae* (1671) and on, showing how Leibniz's approach developed from a strictly mechanistic approach to one that, from the late 1670s and on, insisted that laws of optics could be derived from considerations of optimality.

7 Leibniz, 'Essai anagogique dans la recherche des causes *Tentamen anagogicum*' (1695), in *Die philosophischen Schriften*, ed. Karl Immanuel Gerhardt (Berlin: Weidmann, 1890) bd.7, 270–9.

8 Leibniz, *New Essays on Human Understanding*.

9 John Locke and R. S. Woolhouse, *An Essay Concerning Human Understanding* (London and New York: Penguin Books, 1997).

10 See Locke, Book I, *An Essay*, 55–91. As Remnant and Bennett describe, Leibniz's exposition of Locke's ideas does not 'fairly' present Locke's position; rather, 'instead of two people seriously arguing, we have a mechanical spokesman for Locke (Philalethes) who dutifully serves up portions of the *Essay* so that Leibniz's spokesman (Theophilus) can pass judgment on them'; see 'Introduction', Leibniz, *New Essays*, x.

11 'Involuntary thoughts come to us partly from without, through objects' affecting our senses, and partly from within, as a result of the (often undetectable) traces left behind by earlier perceptions, which continue to operate and mingle with new ones. We are passive in this respect; and even when we are awake we are visited by images – which I take to include representations not only of shapes but also of sounds and other sensible qualities which come to us unbidden, as in dreams.' Leibniz, *New Essays*, 178.

12 Leibniz, *New Essays*, 272.

13 Ibid, 51, 445.

14 Ibid, 488.

15 Ibid, 404.

16 Ibid, 272.

17 Ibid, 404.
18 Ibid, 51.
19 Ibid, 262.
20 Ibid.
21 Plato distinguishes between 'two paired species of the art of image making': 'Eikastics and Phantastics', *The Sophist*, 237a.
22 Ricoeur discusses the association of *eidolon*, image, and *eikon* copy, with *tupos*, imprint through the metaphor of the slab of wax in Plato's *Theaetetus*. See Ricoeur, *Memory, History, Forgetting*, 8–10.
23 Deleuze comments on imagination in Kant in *Difference and Repetition*, 187.
24 See Benedict de Spinoza, 'Letter LII. Spinoza to Leibniz', in *On the Improvement of the Understanding*, trans. R. H. M. Elwes, Dover Books on Western Philosophy (New York: Dover Publications, 1970), 372–3.
25 See Andersen, *The Geometry of an Art: The History of the Mathematical Theory of Perspective from Alberti to Monge*, Sources and Studies in the History of Mathematics and Physical Sciences (New York; London: Springer, 2007), xx. For another important historical and philosophical account of perspective, see Hubert Damisch, *The Origin of Perspective*, 3. print (Cambridge, MA: MIT Press, 2000).
26 Leibniz, *Monadology*, 57, in *Philosophical Essays*, 220.
27 Andersen, *The Geometry of an Art*, 402.
28 See Leibniz, *New Essays*, 135; see also Debuiche, 'L'expression Leibnizienne', 415. Leibniz took some notes on Pascal's conic sections, see Costabel, Pierre. 'Traduction française des notes de Leibniz sur les "Coniques" de Pascal'. In Taton, *L'œuvre scientifique de Pascal*, 90–101, quoted in Debuiche, 'L'expression Leibnizienne', 419.
29 As Remnant and Bennett mention, by Leibniz's time, Desargues' work was unattainable and Leibniz was probably familiar with Desargues' work through A. Bosse's *La maniere universelle de M. Desargues pour pratiquer la perspective* (1648). See Leibniz, 'Notes', *New Essays*, ix.
30 Leibniz, *New Essays*, 135.
31 Ibid.
32 See Serres, *Le système de Leibniz*, 160.
33 Ibid, 165.
34 See Debuiche,'L'expression Leibnizienne', 421.
35 Leibniz, *Theodicy*, 357.
36 See Serres, *Le système de Leibniz*, 159.
37 Ibid, 161.
38 Leibniz, Lettre à des Bosses du 5 fevrier 1712, quoted in Serres, *Le système de Leibniz*, 153.

39 Leibniz, *Theodicy*, 147.

40 As mentioned, Leibniz was actively engaged in optical research. His use of the 'living mirror' metaphor must be contextualized through scientific developments in lens and glass making and the usage of mirrors during his times. As Melchoir-Bonnet describes, while mirrors were in use from antiquity, early mirrors were made of metal and the early glass mirrors, developed during the Middle Ages, were often opaque. From the second half of the fifteenth century glass makers from Morano developed techniques for producing glass so pure and transparent that it was termed 'crystalline', which later spread to France and Germany. However, glass mirrors remained rare and luxurious objects of small size, their surfaces usually concave or convex, until the eighteenth century. See Sabine Melchior-Bonnet, 'The Secret of Venice', in *The Mirror: A History* (London and New York: Routledge, 2014), 9–34. For a general study of optical evolutions in the seventeenth century contributing to the development of devices that newly delineated the boundaries of vision, such as the telescope and the microscope, see A. I. Sabra, *Theories of Light: From Descartes to Newton*, 1 ed. (Cambridge: Cambridge University Press, 1981). See also Damisch, who shows how mirrors were used by Filippo Brunelleschi in the invention of central perspective, in 'The Tradition', *The Origin of Perspective*, 58–72.

41 For example by Chen Moris who links between Kepler's optics and Shakespeare's use of the mirror metaphor, in 'The Quality of Nothing: Shakespearean Mirrors and Kepler's Visual Economy of Science'; see Ofer Gal and Raz Chen-Morris, eds., *Science in the Age of Baroque*, Archives Internationales d'histoire Des Idées = International Archives of the History of Ideas 208 (Dordrecht; New York: Springer, 2013).

42 Most notably in 'Principles of Nature and Grace'. Wilson and Nachtomy have shown that, contrary to the common perception that biology emerged as a full-fledged science in the eighteenth and nineteenth centuries, Leibniz's had a highly defined concept of life. Nachtomy demonstrates this concept is still influential today by creating an analogy between Leibniz's notion of the substance's internal program of action and twentieth-century biologist François Jacob's notion of a genetic program. See Catherine Wilson, 'Leibniz and the Logic of Life', *Revue internationale de philosophie* 48, no. 188 (1994): 237–53, Ohad Nachtomy, 'Leibniz and "The Logic of Life"', *Studia Leibnitiana*, Bd. 41, H. 1 (2009): 1–20.

43 G.W Leibniz, Letter to Samuel Masson (1716), *Philosophical Essays*, 227. Leibniz may have been referring to Plato, who describes the soul as mirror of divinity, for example in *First Alcibiades*, 133c: 'Shall we not say, therefore, that as mirrors are clearer, purer, and more splendid than that which is analogous to a mirror in

the eye, in like manner God is purer and more splendid than that which is best in our soul?'

44 Leibniz, 'Principles of Nature and Grace', in *Monadology and Other Philosophical Writings*, 409.

45 See Ohad Nachtomy, *Living Mirrors: Infinity, Unity, and Life in Leibniz's Philosophy* (New York, NY: Oxford University Press, 2019), 134. For a reconstruction of the fragment see Leibniz and Frédéric de Buzon, 'Double infinité chez Pascal et Monade: Essai de reconstitution de deux états du texte', Presses Universitaires de France, *Les Études philosophiques* 4, no. 95 (2010): 549–6.

46 See de Buzon, ibid, 547.

47 See Rey, 'L'ambivalence de la notion d'action', 47–66.

48 'The union of soul with body, and indeed the operation of one substance upon another, consists only in the perfect mutual accord of substances, definitely established through the order of their first creation, in virtue of which each substance, following its own laws, agrees with the rest, meeting their demands; and the operations of the one thus follow or accompany the operations or change of the other.' Leibniz, Letter to Arnauld (1690), quoted in *Monadology and Other Philosophical Writings*, 106.

49 Leibniz, 'Principles of Nature and Grace' 13, in *Philosophical Essays*, 211.

50 Leibniz, *Monadology*, 83, 266.

51 Ibid, 12, 223.

52 As Samuel Weber writes, 'The detail is that aspect of the monad that makes perception possible and with it, differentiation … [It] marks the aspect of the monad where consciousness converges with unconsciousness.' Weber, 'God and the Devil – in Detail', in *Benjamin's -abilities*, 246.

53 Leibniz, Letter to Arnauld from April 1687, quoted in Deleuze, *The Fold*, 85.

54 Deleuze, ibid, 86.

55 Deleuze, *The Fold*, 99/*Le Pli*, 132.

56 Ibid, 120/90.

57 Ibid, 115.

58 Leibniz, *New Essays*, 111; book 2, chap 1, 10.

59 Deleuze, *The Fold*, 118.

60 Ibid, 89.

61 Whitehead, *Process and Reality*, 129, quoted in Deleuze, 154.

62 Deleuze, *The Fold*, 86.

63 As Deleuze writes, 'The finite infinite present in the finite self is exactly the position of the equilibrium, or disequilibrium, of the Baroque.' *The Fold*, 89 (modified)/*Le Pli*, 119.

64 The Claude glass, usually a small, black convex mirror, was named after seventeenth-century landscape painter Claude Lorrain and used by artists to add

tonal values to the reflection of a view. Arnaud Maillet traces its wider uses from occult practices to photography and film in *The Claude Glass: Use and Meaning of the Black Mirror in Western Art*, First paperback ed. (New York: Zone Books, 2009).

65 See Benjamin, 'Zur Moral und Anthropologie', *GS* 6, 54–89.

66 Benjamin, 'Surrealism', *SW* 2, 217. See Léa Barbisan, *Le corps en exil: Walter Benjamin, penser le corps*, Bibliothèque allemande (Paris: Éditions de la Maison des sciences de l'homme, 2020) and 'Eccentric Bodies, From Phenomenology to Marxism – Walter Benjamin's Reflections on Embodiment', *Anthropology and Materialism*, Special Issue 2017 'Discontinuous Infinities': http://journals.openedition.org/am/803; Sigrid Weigel, *Body-and Image-Space: Re-Reading Walter Benjamin*. Warwick Studies in European Philosophy. London; New York: Routledge, 1996.

67 See Barbisan, 'Eccentric Bodies, From Phenomenology to Marxism'.

68 Benjamin, 'Wahrnehmung und Leib', *GS6*, 67 (my translation).

69 Benjamin, 'Outline of the Psychophysical Problem', *SW* 1, 395 modified /*GS* 6, 80.

70 Ibid, 396/81.

71 Benjamin, *SW* 1, 92.

72 Benjamin, 'The Work of Art in the Age of Its Technological Reproducibility: Second Version', *SW* 3, 113. In a footnote, Benjamin writes that 'the significance of beautiful semblance [*schöner Schein*] is rooted in the age of auratic perception that is now coming to an end. The aesthetic theory of that era was most fully articulated by Hegel'. See fn22, 127.

73 As Benjamin writes, 'the oldest form of imitation had only one material to work with: the body of the mime himself'. Ibid.

74 Ibid, 113.

75 Ibid, 109.

76 Yvonne Zimmermann for example reflects on how in videoconferencing, the mirror metaphor, long used by film theorists to describe the screen, becomes a reality, resulting in a distancing as well as an encounter with the self. See 'Videoconferencing and the Uncanny Encounter with Oneself: Self-Reflexivity as Self-Monitoring 2.0', Philipp Dominik Keidl et al., eds., *Pandemic Media: Preliminary Notes toward an Inventory*, Configurations of Film (Lüneburg: meson press, 2020). See also Shane Denson, who discusses what he calls a 'discorrelation' between machine-generated images in our contemporary, new media saturated world and human perception, incurring a shift in the relationship between perception and its objects. See *Discorrelated Images* (Durham: Duke University Press, 2020).

77 'Thou shalt not make unto thee any graven image' – this serves not just to avert the worship of idols. With incomparable emphasis, the prohibition of the *representation of the body* safeguards against the illusion that the sphere in which

the moral essence of humanity is perceptible can be reproduced. Benjamin, *Origin of the German Trauerspiel,* 98, my emphasis.

78 Benjamin, 'The Work of Art in the Age of Its Technological Reproducibility: Second Version', 127–8, fn 22.

79 Adorno, 'The Idea of Natural History'.

80 Benjamin is alluding to György Lukács' concept of 'Second Nature', developed in *The Theory of the Novel: A Historico-Philosophical Essay on the Forms of Great Epic Literature*, 12. print (Cambridge, MA: The MIT Press, 1994). The concept is used in the First Version of Benjamin's Artwork essay but not the Second, in which he opts for 'Other Nature'. For a reading that analyses the variations between the different versions in relation to Benjamin's conception of nature and use of Lukács' concept, see Andreas Becker: Von der zweiten zur anderen Natur. Ein Motiv in Walter Benjamins Kunstwerk-Aufsatz. In: Thomas Barth, Christian Betzer, Jens Eder u.a. (Hg.), *Mediale Spielräume* (Marburg: Schüren, 2005 (Film- und Fernsehwissenschaftliches Kolloquium 17), 177–86, https://doi.org/10.25969/mediarep/14431. On the different versions of the manuscript, see Eiland and Jennings, 'Notes', Benjamin, ibid, 122.

81 Ibid, 107–8 and 127–8, fn. 22.

82 Ibid, 108.

83 Benjamin's use of the term 'mankind' implies a gendered distinction between the ancient, feminine community and communality with nature, and the modern, masculine society that is alienated from nature, which I cannot explore here in depth.

84 To the Planetarium, *One Way Street*, 94–5.

85 Pauline Phemister, 'Leibniz and Ecology', *History of Philosophy Quarterly* 18, no. 3 (July, 2001): 246.

86 Here we differ from McBride, who describes Benjamin's conception of montage as primarily inspired by cinematic technology. She further suggests that that Benjamin's montage inaugurates a mode of narrative that responds to the alienation of modern urban life. See Patrizia C. McBride, *The Chatter of the Visible: Montage and Narrative in Weimar Germany* (Ann Arbor: University of Michigan Press, 2016), 41–2.

87 *SW* 2, 1, 301.

88 *SW* 4, 406, modified.

89 Benjamin, *Arcades Project*, [N10a,3], 475.

90 See Benjamin, *Arcades Project*, [N9,7], 473 and [N10a,3], 475.

91 'Truth [...] is bound to a nucleus of time hidden in the knower and the known alike', Benjamin, *Arcades Project*, [N3,2], 463.

92 Ibid, [N10,3], 475/*GS*5, 594.

93 As she writes: 'The principle of construction is that of montage, whereby the image's ideational elements remain unreconciled', Susan Buck-Morss, *The*

Dialectics of Seeing: Walter Benjamin and the Arcades Project (Cambridge, MA: MIT Press, 1999), 67.

94 Benjamin, *Arcades Project*, [N1a,8], 460.

95 Benjamin, 'On the Concept of History', 389–400.

96 In 'Reduktion und Montage', Detlev Schöttker identifies montage in Benjamin's writing with construction, arguing for the link between Benjamin's interest in constructivist movements and his development of the concept of montage. This leads him to wonder why on one hand Benjamin positions 'construction' against 'epic theatre', while on the other arguing that epic theatre is montage-like. Schöttker explains this by suggesting that over the years Brecht appropriated the idea of historical construction from Benjamin, employing it in his own writing and, therefore, Benjamin later references to Brecht's work link epic theatre together with montage. While I agree with the link between montage and constructivism, and find Schöttker's observation of shifts over time in Brecht's approach to construction convincing, the apparent contradiction is avoided if one views montage as a complex, evolving concept in Benjamin's writing that includes moments of both construction and interruption. *Global Benjamin: Internationaler Walter-Benjamin-Kongress 1992* (München: Wilhelm Fink Verlag, 1999), bd.2, 772–3.

97 The impact of Brecht's theorization of the *Verfremdungseffekt* on Benjamin's conceptualization of montage as an interruptive procedure is apparent in Benjamin's 'The Author as Producer' and his essays on Brecht's epic theatre; the *procedure* of montage induces spectators to judgment and political engagement, as theorized by Brecht.

98 Benjamin, *Arcades Project* [N2,6], 461/*GS* 5, 575. See also Noa Levin, 'Montage Mahagonny: Walter Benjamin and Bertolt Brecht's Theatre of Interruptions', *Material und Begriff. Arbeitsverfahren und theoretische Beziehungen Walter Benjamins* (Hamburg: Argument, 2019), 145–59.

99 Adorno, 'A portrait of Walter Benjamin', *Prisms*, 238.

100 André Breton, 'Manifesto of Surrealism' 1924, *Manifestoes of Surrealism*. 1. ed. as an Ann Arbor paperback. Ann Arbor Paperbacks 182. Ann Arbor, Michigan: University of Michigan Press, 1972, 41.

101 Benjamin, *Origin of the German Trauerspiel*, 12.

102 See Lev Kuleshov's 'Kuleshov Effect' from 1921, in which he demonstrated the way in which viewers interpret the juxtaposition between shots to bear different emotional meanings: https://www.youtube.com/watch?v=_gGl3LJ7vHc.

103 *Translator's Foreword* in Benjamin, *The Arcades Project*, xi.

104 Benjamin, 'The Storyteller', 153.

105 In an early fragment Benjamin replaces cause and effect with the category of debt [*Schuld*], as a determining relation in history: 'the ultimate category world-history

uses in order to guarantee the insightibility of events is debt/guilt [*Schuld*]. Every world-historical moment is indebted to other moments and other moments are indebted to it. Cause and effect can never be decisive categories for the structure of world-history, for they cannot determine a totality.' 'Zur Geschichtesphilosophie, Historik und Politik' *GS*6, 92 (my translation). The word *Schuld* refers both to capitalism as the driving political framework of history and to Freud's theory of guilt and repression.

106 Adorno writes that 'In Germany the essay provokes resistance because it is reminiscent of the intellectual tradition that from the time of an unsuccessful and lukewarm Enlightenment, since Leibniz's day, all the way to the present has never really emerged, not even under the conditions of formal freedom; the German Enlightenment was always ready to proclaim, as its essential concern, subordination under whatever high courts', 'The Essay as Form', 152.

107 Ibid, 166, 162.

108 Buck-Morss, 'Aesthetics and Anaesthetics', 6.

109 As Terry Eagleton writes, 'it is as though philosophy suddenly wakes up to the fact that there is a dense, swarming territory beyond its own mental enclave which threatens to fall utterly outside its sway. That territory is nothing less than our whole sensate life together – the business of affections and aversions, of how the world strikes the body on its sensory surfaces … [T]he aesthetic concerns this most gross and palpable dimension of the human, which post-Cartesian philosophy, in some curious lapse of attention, has somehow managed to overlook', *The Ideology of the Aesthetic* (Oxford: Blackwell, 1990), 13.

110 Buck-Morss does, however, discuss the significance of images to Benjamin in *The Dialectics of Seeing*, where she recounts how the *Arcades Project* was at a certain point planned to include images, as Benjamin wrote in a letter to Gretel Karplus. See Buck-Morss, *The Dialectics of Seeing*, 71.

111 Ibid, 31.

112 Deleuze, *The Fold*, 56.

113 Ibid, 20.

114 Deleuze, *Cours Vincennes*, Cours du 15/04/1980.

115 Deleuze, *The Fold*, 20/*Le Pli*, 27.

116 Ibid.

117 Ibid, 21/30. Deleuze reuses the same quote to describe perspectivism, this time without attributing it to Serres, in *What Is Philosophy?*, 130.

118 Whitehead, *Process and Reality*, quoted in Gentry, 'The Subject in Whitehead's Philosophy', 222.

119 Deleuze, *The Fold*, 32.

120 Ibid, 21.

121 Ibid, 28. Here his reading of Leibniz differs from Benjamin's, which, as we have seen, highlights the distinct points of view of the monads as enacting breaks in the continuity.

122 Following a description of the challenges of separating mannerism from baroque, Richard Sayce suggests the following periodization: 'Renaissance se charchterise par l'espace statique, la composition équilibrée, la lumière harmonieusement répandue; le maniérisme par le *Raumflucht* …, la tension dans le composition, la lumière glissant; le baroque par la profondeur dans l'espace, la composition qui s'épanouit, la lumière inondant'. 'Maniérisme et périodisation, quelque réflexions général', in *Renaissance, Maniérisme, Baroque* (Paris: Library Philosophique J.Vrin, 1972), 48.

123 Max Dvořák was one of the first post-war art historians to reconsider Mannerism, describing El Greco as a proto avant-garde artist in a lecture given in 1920. See 'Über Greco und den Manierismus', in *Kunstgeschichte als Geistesgeschichte. Studien zur abendländischen Kunstentwicklung*, hg. v. Karl Maria Swoboda und Johannes Wilde (München: Piper, 1924), 259–76. Benjamin refers to this text in 'Rigorous Study of Art' writing that it 'demonstrated what historical insights can be gleaned from a spiritualist distortion of the now empty schemata of the pure classical period. All this was overlooked by a rigidly periodic universal history' (Wölfflin's). See fn.22, 89.

124 See Walter Friedlaender: 'Thus arises a new beauty, no longer resting on real forms measurable by the model or on forms idealised on this basis, but rather on inner artistic reworkings on the basis of harmonic or rhythmical requirements', in 'The Anticlassical style', *Mannerism and Anti-Mannerism in Italian Painting*, 9. print (New York: Schocken Books, 1985), 8.

125 Deleuze, *The Fold*, 36–7 (modified)/*Le Pli* 51.

126 'The usual formula of the Baroque still life is: drapery, producing folds of air or heavy clouds; a tablecloth with maritime or fluvial folds, jewellery that burns with folds of fire … the painting is so packed with folds that their results a sort of schizophrenic "stuffing" … it seems that this ambition of covering the canvas with folds is discovered again in modern art, with the "all-over" fold.' Deleuze, *The Fold*, 122–3.

127 Deleuze, *The Fold*, 34; El Greco, *Christ in the Olive Garden*, c.1600, oil on canvas, 138 x 92 cm, Palais des Beaux-Arts de Lille, Lille, France.

128 Ibid, 122.

129 Ibid, 53.

130 Deleuze, *The Fold*, 19.

131 Ibid.

132 See the 'Non Standard Architectures' exhibition at Centre Pompidou, Paris (December 2003–March 2004), on the use of numerical tools in architecture.

133 See Mario Carpo, 'Introduction', in Bernard Cache, *Projectiles*, Architecture Words 6 (London: Architectural Association, 2011), Bernard Cache and Christian Girard, 'Objectile: The Pursuit of Philosophy by Other Means?', in *Deleuze and Architecture*, ed. Hélène Frichot, Deleuze Connections (Edinburgh: Edinburgh University Press, 2013), 96–110 and 'Deleuze and the Story of the Superfold', ibid, 79–95.

134 Bernard Cache, *Earth Moves: The Furnishing of Territories*, 10. [pr.], Writing Architecture (Cambridge, MA: MIT Press, 2010), 95.

135 See Mario Carpo, *Beyond Digital* and 'A short but believable history of the digital turn in architecture': https://www.e-flux.com/architecture/chronograms/528659/a-short-but-believable-history-of-the-digital-turn-in-architecture.

136 See Martin Davis, *The Universal Computer: The Road from Leibniz to Turing* (New York: Norton, 2000).

137 Deleuze, *The Fold*, 19.

138 See Bernard Cache, *Earth Moves*, 17–18 and Deleuze, *The Fold*, 19.

139 Ibid, 21.

140 Ibid, 56.

141 Ibid, 32.

142 Ibid, 56–7, 31.

143 Ibid, 32 modified/45.

144 Ibid, 188.

145 Deleuze, ibid, 58.

146 Deleuze, ibid, 69, 152 n23. In *Cinema 2*, the 'mannerist' conversations of schizophrenics are depicted as creating distance, in a similar way to the distance between monads in Leibniz's perspectivism. As he writes: 'Psychiatrists have studied the conversation of schizophrenics, with its mannerisms, its interactional bringing closer and putting at a distance, but all conversation is schizophrenic, conversation is a model of schizophrenia, not the other way round.' *Cinema 2*, 230.

147 Benjamin, *Origin of the German Trauerspiel*, 56.

148 Benjamin quotes from Daniel Caspar von Lohenstein's *Sophonisbe*: 'through Disalces, a messenger, Masinissa sends poison to Sophonisbe, so she can elude Roman imprisonment: "Be quiet, go, Disalces, my patience do not strain! / But halt! I feel I'm done for, going numb, and trembling. / Go on! No time to doubt. Suspense I'm not dissembling. / Forgive! Oh, look, my heart and eye both break! / Go on! For the end no other course can take."' *Origin*, 56–7. See Daniel Casper von Lohenstein, *Sophonisbe: Trauerspiel*, ed. Rolf Tarot, Bibliographisch aktualisierte Ausgabe, Reclams Universal-Bibliothek, Nr. 14234 (Ditzingen: Reclam, 2021).

Chapter 5

1 Benjamin uses *Abbild* to designate a facsimile or reproduction in the 'Artwork' essay: 'the reproduction [*Reproduktion*], as offered by illustrated magazines and newsreels, differs unmistakably from the image' *SW* 3, 105.
2 Benjamin, *Berlin Childhood around 1900*, in *SW* 3, 344–413.
3 See Rita Copeland and Peter T. Struck, eds., *The Cambridge Companion to Allegory*, Cambridge Companions to Literature (Cambridge, UK; New York: Cambridge University Press, 2010), 3.
4 Even talented authors such as William Butler Yeats and Arthur Schopenhauer, Benjamin writes, were mistaken about the allegory. See *Origin*, 169.
5 Benjamin, *Origin of the German Trauerspiel*, 169, modified /*GS*1, 339.
6 Ibid, 188.
7 Benjamin, *Origin of the German Trauerspiel*, 189 modified/*GS*1, 354.
8 Ibid, 185.
9 Görres, cited in Creuzer, *Symbolik und Mythologie der alten Völker, besonders der Griechen* (Leipzig, Darmstadt, 1810–1819), 147–8, cited in ibid, 173.
10 Ibid, Benjamin, 167.
11 'Die Geschichte wandert in den Schauplatz hinein', ibid, 81.
12 Ibid, 188.
13 Adorno, 'The Idea of Natural History'.
14 Benjamin, *Origin of the German Trauerspiel*, 184.
15 'Baudelaire's allegory bears the traces of the violence that was necessary to demolish the harmonious façade of the world that surrounded him' [J55a, 3], *Arcades Project*, 329. See also ibid, [J53, 3], 324 and 'Central Park', *SW4*, 169, 174.
16 For more on modern and Baroque allegories see Buck Morss, 'Historical Nature: Ruin', *The Dialectics of Seeing*, 159–201.
17 See for example Benjamin Buchloh, 'Allegorical Procedures: Appropriation and Montage in Contemporary Art', in *Art After Conceptual Art* (Cambridge: MIT Press, 2006), 27–51; Peter Bürger, 'The Avant-Gardiste Work of Art', in *Theory of the Avant-Garde*, Theory and History of Literature 4 (Manchester: Manchester University Press [u.a.], 1984), 55–82. See also Jack Wong, 'Remapping the Constellation of Walter Benjamin's Allegorical Method', *American, British and Canadian Studies* 25, no. 1 (3916): 37–59. https://doi.org/10.1515/abcsj-2015-0007.
18 Bürger, ibid, 68.
19 Deleuze, *The Fold*, 123.
20 Ibid, 125 (modified)/*Le Pli*, 170–1.
21 Ibid, 121.
22 Deleuze, *The Fold*, 125 modified/*Le Pli*, 171.
23 Ibid, 125.

24 Ibid.
25 Deleuze refers to Serres, *Le système de Leibniz*, 2, 653–7.
26 Ibid, 126.
27 See Jon R. Snyder, 'Art and Truth in Baroque Italy, or the Case of Emanuele Tesauro's Il cannocchiale aristoteli', *MLN* 131, no. 1 (2016): 74–96 for an analysis of the book and its Frontispiece.
28 Deleuze, *The Fold*, 126.
29 Deleuze, ibid, 137.
30 Benjamin, 'Surrealism', *SW*2, 208.
31 Ibid.
32 Ibid, 209.
33 Benjamin, 'One-Way Street', *SW* 1, 444/*GS*4, 'Tankstelle', 83.
34 Benjamin, *SW* 1, 444/*GS* 5, 87.
35 Sergei Tretyakov, 'The Writer and the Socialist Village', *October* 118 (Fall 2006): 63–70. Benjamin refers to his ideas in 'The Author as Producer', *SW* 2, 770.
36 Benjamin, *SW* 1, 456/'Vereidigter Bücherrevisor', in *Einbahnstraße*, *GS*4, 102–4.
37 Benjamin, *Origin of the German Trauerspiel*, 175–7.
38 Louis Aragon, *Treatise on Style*, trans. Alyson Waters (Lincoln, NE: University of Nebraska Press, 1991).
39 See Alyson Waters, *Treatise on Style*, 'Translator's Introduction', xiii.
40 Benjamin may have been inspired by the following passage: 'Literary images are in fact the vehicles of humour, and by proportional reciprocity, humour is what gives an image its force. Compare two images taken at random and you will be reduced to dust. Which also explains the way they age, because humour is assigned to an image only for a short time, and as soon as it has remounted its motorcycle, the wall begins to crumble. This is the basis of the idea of poetic novelty, about which there has recently been so much commotion … poetry is by nature stormy, and every image should produce a cataclysm.' Aragon, *Treatise on Style*, 69.
41 Benjamin, 'Surrealism', *SW*2, 216.
42 Benjamin's distinction between image and metaphor is comparable to that between affinity and analogy noted in Chapter 1.
43 Deleuze, Gilles, Guattari, Félix. *Kafka: Toward a Minor Literature*. 9. print. Theory and History of Literature 30. Minneapolis: University of Minnesota Press, 2008.
44 In his essay on Beckett, 'The Exhausted', Deleuze comes even closer to Benjamin, denoting the image as 'intensity': 'The image is what dies away, wastes away, a fall. It is a pure intensity, which defines itself as such through its height-level above zero, which is only described in falling.' Here 'intensity' equals ephemerality, the image comprises the process of its own dissolution. 'The Exhausted', trans. Anthony Uhlmann, *Substance* 24, No. 3, no. 78 (1995): 19.
45 Ibid.
46 Benjamin, 'Surrealism', *SW*2, 217 (modified)/*GS* 2, 309.

47 Ibid, 217–18, (modified)/*GS* 2, 310.
48 Ibid, 207/297.
49 In 'To the Planetarium', the last *Denkbild* in *One-Way Street*, Benjamin similarly describes technology as the new organizing principle of *physis* – the relation between mankind and the world – that replaces nations and families. Benjamin, *SW*1, 486–7.
50 Deleuze, 'Preface to the English Edition', *Difference and Repetition*, xvi–xvii.
51 Deleuze, Logic of sense, 253/*Logique du sens*, 292.
52 Ibid.
53 Ibid, 296.
54 Deleuze, *Logic of Sense*, 256.
55 Ibid.
56 Ibid, 262. Deleuze's conception of the simulacrum differs from the more renowned conception of Baudrillard's simulacra as copies that never had an original.
57 Ibid, 258.
58 Ibid, 132.
59 Deleuze and Guattari, *What Is Philosophy?*, 37.
60 Ibid, 40.
61 Deleuze, *Difference and Repetition*, 1, modified/*Différence et répétition*, xix.
62 Ibid, 68–9 (modified)/94.
63 Deleuze, *Logic of Sense*, 260.
64 Deleuze and Guattari, *What Is Philosophy?*, 204.
65 For a discussion on the ways in which Deleuze and Guattari were informed by quantum field theory see Arkady Plotinsky, 'Chaosmologies: Quantum Field Theory, Chaos and Thought in Deleuze and Guattari's What Is Philosophy?', *Paragraph* 29, no. 2, Deleuze and Science (July 2006): 40–56.
66 Deleuze, *Cinema 1*, xviii. In Deleuze's words, 'Images, in cinema, are signs. Signs are images seen from the viewpoint of their composition and generation … Cinema has given rise to its own particular signs, whose classification is specific to cinema, but once it produces them they turn up elsewhere, and the world starts "turning cinematic"'. Deleuze, *Negotiations*, 65.
67 See for example Deleuze, *Cinema 2*, 24.
68 Deleuze, *Cinema 1*, 3.
69 Henri Bergson, *Matter and Memory* (New York: Zone Books, 1988).
70 Ibid, 17–18.
71 Deleuze, 'Glossary', *Cinema 1*, 241.
72 Deleuze, *Cinema 1*, xii.
73 Deleuze, *Cinema 2*, 127/*Cinéma* 2, 166–7.
74 Deleuze, *Cinema 2*, 59. The 'point of indiscernibility' recalls Leibniz's 'method of indetermination' described by Serres, *Le système de Leibniz*, 221.
75 Ibid, 44–7, 68.

76 Ibid, 74/100.

77 Ibid, 81/109. Deleuze quotes Jean Ricardou, *Le Nouveau Roman*, Points Littérature 212 (Paris: Ed. du Seuil, 1990), 73.

78 Ibid, 70.

79 Ibid, 76.

80 Deleuze, *Cinema 2*, 80–1.

81 Bergson, *Matter and Memory*, 38/Henri Bergson, *Matière et mémoire. Essai sur la relation du corps à l'esprit*, Une édition électronique réalisée à partir du livre de c Première édition: 1939 (Paris: Les Presses universitaires de France, 1965), 22.

82 Deleuze, *Cinema 2*, 81 modified/*Cinéma 2*, 108–9.

83 Ibid, 81/109.

84 See Deleuze, Gilles, *Bergsonism* (New York: Zone Books, 1988), 52.

85 Benjamin, *Arcades Project*, [N2a, 3]; Deleuze, *Cinema 2*, 25.

86 See for example Eli Friedlander, *Walter Benjamin: A Philosophical Portrait* (Cambridge, MA: Harvard University Press, 2012), 37.

87 Benjamin, *Arcades Project*, [N2a, 3], 462.

88 See Deleuze, *Cinema 2*, 81.

89 Peter Osborne writes that Benjamin's dialectic images are 'constellations of the "then" and "now", which, in the hermetic enclosure of their internal relations, mirror the structure of history as a whole, from the standpoint of its end'. 'Modernity, Eternity, Tradition', Peter Osborne, *The Politics of Time: Modernity and Avant-Garde* (London; New York: Verso, 1995), 145.

90 Benjamin, *Arcades Project*, [N3, 1], 463.

91 See Deleuze, *Cinema 2*, 24.

92 Benjamin, *Arcades Project*, [N3, 1], 463.

93 Ibid, [N2, 6], 461.

94 See Thesis XVIII, *SW*4, 396, in which 'now time' is described as figural.

95 Benjamin, *Arcades Project*, [N7a1, 2], 470 modified, thesis XVII, 396. As Löwy describes, the debate between 'utopian (messianic or revolutionary) images' and 'formless progressive tendencies' is one which is posed as early as Benjamin's 'The Life of Students' (1914) as well as in his dissertation on 'The Concept of Criticism in German Romanticism' (1919).

96 Benjamin, 'On the Concept of History', *SW4*, 389–400. The title is sometimes translated as 'Theses on the Philosophy of History'. See also 'Paralipomena to "On the Concept of History"', ibid, 401–11.

97 Löwy, *The Fire Alarm*, 52.

98 Benjamin, *On the Concept of History*, Thesis XIV, *SW*4, 395 modified.

99 Ibid, Thesis XVII, 396.

100 '"Die kümmerlichen fünf Jahrzehntausende des homo sapiens", sagt ein neuerer Biologe, "stellen im Verhältnis zur Geschichte des organischen Lebens auf der Erde etwas wie zwei Sekunden am Schluß eines Tag von vierundzwanzig Stunden

dar. Die Geschichte der zivilisierten Menschheit würde, in diesem Maßst [~~xx~~] abzt ~~damit der Wahrheit Rechnung~~ eingetragen, ein fünftel der letzten Sekunde der letzten Stunde [~~x~~] füllen'". Benjamin, *Über den Begriff der Geschichte*, ed. Gérard Raulet, 1. Aufl, *Werke Und Nachlass / Walter Benjamin*, Bd. 19 (Berlin: Suhrkamp, 2010), 28. Benjamin seems to have been aware of geological time scales and the relatively small scale of the history of humankind. This conjunction of time scales is something that Anthropocene theorists such as Chakrabarty and Zalasiewicz have recently called attention to in relation to the climate crisis. See Dipesh Chakrabarty, *The Climate of History in a Planetary Age* (Chicago; London: The University of Chicago Press, 2021), 15, and Jan Zalasiewicz et al., eds., *The Anthropocene as a Geological Time Unit: A Guide to the Scientific Evidence and Current Debate* (Cambridge New York, NY Port Melbourne New Delhi Singapore: Cambridge University Press, 2019).

101 Benjamin, ibid. For an account of the different existing versions of the 'Theses' and the tangled history of their publication, see ibid, 'Entstehungs- und Publikationgeschichte', 161–208.

102 Benjamin, *SW*1, 168/*GS*1, 92. See Hamacher, *Now Time*, 49 and Löwy, 'Introduction: Romanticism, Messianism and Marxism in Walter Benjamin's Philosophy of History', *Fire Alarm*.

103 Benjamin, *SW*3, 305/*GS*6, 203. Some, however, date the text to 1937–38.

104 Ibid.

105 Ibid.

106 Ibid, *schwache* italicized in original.

107 Löwy, *The Fire Alarm*, 79.

108 Benjamin, *SW*4, 390, 396 modified II, XVII, *GS*1, 694, 703.

109 See Löwy, *The Fire Alarm*, 72, However, Lotze's critique of progress is by no means Leibnizian.

110 Benjamin, *SW* 4, 389–90.

111 Benjamin, *Arcades Project* [N13a, 3], 479.

112 See Löwy, *The Fire Alarm*, 83.

113 Benjamin, *Arcades Project* [N13, 2], 478. As discussed in Chapter 3, Leibniz also used the image of a spiral to describe history in 'Demonstrationes de Universo immenso aeternoque; de Mundis et aevis, deque rerum longiquarum et futurarum statu, c. February–March 1701', *De l'horizon de la doctrine humaine: 1693*, 58–9.

114 Ibid, [N8, 1], 471.

115 Horkheimer had sent Benjamin his essay on Bergson, titled 'On Bergson's Metaphysics of Time'.

116 Max Horkheimer, 'On Bergson's Metaphysics of Time', *Radical Philosophy* 131 (2005): 10.

117 Deleuze, *Cinema 2*, 41 (modified)/*Cinéma 2*, 59.

118 Ibid.
119 Deleuze, *Cinema 1*, 13.
120 Gilles Deleuze, 'Vérité et temps' *Cours Vincennes* cours du 24/04/1984, transcription: lucie Picandet, http://www2.univ-paris8.fr/deleuze/article.php3?id_article=346.
121 Deleuze, *Cinema 2*, 42.
122 Ibid/Cinéma 2, 60.
123 Deleuze, *Difference and Repetition*, 88.
124 See Deleuze, *Cinema 2*, 130–1. See also 'Incompossibility, Individuality, Liberty', *The Fold*, 59–75. For Leibniz on his concept of 'compossibility' see 'Letters to Louis Bourguet', *Philosophical Papers and Letters*, 661–2.
125 'This is Borges's reply to Leibniz: the straight line as force of time, as labyrinth of time, is also the line which forks and keeps on forking, passing through *incompossible presents* and *not-necessarily true pasts*'. Deleuze, *Cinema 2*, 131.
126 Ibid, 132.
127 Deleuze, *Cinema 2*, 51/*Cinéma 2*, 57.
128 Deleuze, *Nietzsche and Philosophy*, 90.
129 Benjamin, *Origin of the German Trauerspiel,* 12; *Arcades Project*, [N3,1], 463. In the Foreword he also refers to the distinction between 'truth content', sought by critique and 'material content', that is sought by commentary, made in the essay 'Goethe's Elective Affinities', SW 1, 297.
130 See Rolf Tiedemann, 'Einleitung des Herausgebers', in *Passagen*, *GS* 5, 30.
131 Benjamin, 'The Work of Art', *SW* 3, 109.
132 See Benjamin, *The Crisis of the Novel*, *SW*2, 299–304; *The Author as Producer*, *SW*2, 674–5.
133 Benjamin, *Arcades Project*, [N1a, 8], 460.
134 As Josh Alvizu suggests, while Leibniz's system is famous for being hierarchized, the concept of isotropy, that forms the basis of Leibniz's concept of expression, according to which magnitudes remain stable through direction or measurement, is perhaps as significant a concept in his thought. See Josh Alvizu, 'Montage Monadology: Leibniz, Benjamin, Modernity' (presentation at *Benjamin and Leibniz: On Expression*, Goldsmith University of London, 27–8 June 2017).
135 Walter Benjamin, *Moscow Diary*, ed. G. Smith (Cambridge, MA: Harvard University Press, 1986), 28, 106, 103; Benjamin, 'On the Present Situation of Russian Film', *SW*2, 12–15.
136 See Benjamin, 'Mickey Mouse', *SW* 2, 545.
137 Benjamin, 'The Work of Art', *SW* 3, 109.
138 Ibid, 126.
139 Ibid. Vsevolod Pudovkin, *Selected Essays*, ed. Richard Taylor (Chicago: University of Chicago Press, 2006), 106.
140 Deleuze, *Cinema 1*, 31.

141 Benjamin, 'The Work of Art', *SW* 3, 113.

142 Moreover, Deleuze highlights that for Leibniz the multiplicity of worlds, or universes, do not all exist in actuality. This is contrary to the current day 'multiverse' argument for the actual existence of infinite universes, for example in David K. Lewis's modal realism as expounded in, *On the Plurality of Worlds* (New Jersey: Wiley-Blackwell, 1986).

143 Deleuze, *The Fold*, 82.

144 Deleuze, *Cinema 2*, 303n5/*Cinéma 2*, 171.

145 Ibid.

146 Leibniz, *Theodicy*, 415, 372.

147 As Schaub suggests, quoting Leibniz: 'for the pyramid had a beginning, but one could not see its end; it had an apex, but no base', *Theodicy*, 416, 373; See Mirjam Schaub, 'Transparenz, opak. Eine Begehung von Leibnizens Kristallpalast am Ende der Theodizee', in Markus Rautzenberg/Andreas Wollsteiner (Hg.), *Hide and Seek*, Das Spiel von Transparenz und Opazität (München: Fink 2010), 143.

148 Laurentius Valla, *Dialogue on Free Will*, trans. C. Trinkaus in *The Renaissance Philosophy of Man*, ed. E. Cassirer et al. (Chicago: University of Chicago Press, 1948).

149 Anicius Manlius Severinus Boethius, *Consolation of Philosophy*, trans. J. C. Relihan (Hackett: Indianapolis, 2001). Written around 524, *Consolation of Philosophy* is a prosimetrum (prose with verse intervals), which unfolds an imaginary dialogue between the imprisoned Boethius and 'Lady philosophy' in which the latter defends rational consolation. It was criticized by Valla and others for its apparent preference of philosophy over the Christian doctrine. See Margaret Cameron, 'Ac pene Stoicus: Valla and Leibniz on "The Consolation of Philosophy"', *History of Philosophy Quarterly* 24, no. 4 (October 2007): 337–54.

150 Leibniz, *Theodicy*, 414, 371.

151 'Thereupon the Goddess led Theodorous into one of the halls of the palace: when he was within, it was no longer a hall, it was a world *Solemque suum, sua sidera norat*'. Leibniz, *Theodicy*, 415, 372.

152 Leibniz, *Theodicy*, ibid.

153 Schaub, 'Transparenz, opak', 147.

154 Benjamin, *Origin of the German Trauerspiel*, 192.

155 See Cameron, 'Ac pene Stoicus', 344. The key role of women, embodied by Lucretia, 'Lady philosophy' and Pallas Athena, in this foundational myth for modernity, is significant yet cannot be dealt with extensively here.

156 Leibniz, ibid, 417, 377.

157 Ibid.

158 Deleuze, *Cinema 2*, 131.

159 Benjamin, *Origin of the German Trauerspiel*, 82.

Conclusion

1 Hito Steyerl, 'In Free Fall: A Thought Experiment on Vertical Perspective', in *The Wretched of the Screen* (Berlin: Sternberg Press, 2012), 12–30.
2 Ibid, 26.
3 See Eyal Weizman, *Hollow Land: Israel's Architecture of Occupation*, New ed. (London: Verso, 2017).
4 Deleuze, *Cinema 2*, 265.
5 In 'Attested Auditor of Books', Benjamin had described the inverse process concerning the written word, in which it had been forced into the 'directorial perpendicular', and that culminated in the three-dimensionality of the card index. Benjamin, *SW* 1, 456.
6 See Leibniz, 'Explication de l'arithmétique binaire, qui se sert des seuls caractères O et I avec des remarques sur son utilité et sur ce qu'elle donne le sens des anciennes figures chinoises de Fohy' (Mémoires de mathématique et de physique de l'Académie royale des sciences, Académie royale des sciences, 1703). https://hal.archives-ouvertes.fr/ads-00104781.
7 See Davis, *The Universal Computer*.
8 See Leibniz, 'Dissertation on the Art of Combinations, 1666 (selection)', *Philosophical Papers and Letters*, 73–84, where Leibniz begins to develop ideas for a universal characteristic.
9 Deleuze, *The Fold*, 38–39/*Le Pli*, 27.
10 Recent developments in language and image processing and machine learning render the link between order and disorder, chaos and probability (as observed by Deleuze) all the more evident. As Steyerl writes, 'visuals created by ml tools are statistical renderings, rather than images of actually existing objects. They shift the focus from photographic indexicality to stochastic discrimination. They no longer refer to facticity, let alone truth, but to probability.' 'Mean Images', NLR, no. 140/141, March/June 2023. See also Sybille Kramer, 'Mind, Symbolism, Formalism: Is Leibniz a Precursor of Artificial Intelligence', *Knowl.Org* 23, no. 2 (1996): 84–7.
11 Voltaire, *Candide*.
12 See Andrew Culp, *Dark Deleuze* (Minneapolis: University of Minnesota Press, 2016).
13 See Russell, *A Critical Exposition of the Philosophy of Leibniz*; Couturat, La logique de Leibniz d'après des documents inédits; Ernst Cassirer, 'Leibniz' System in seinen wissenschaftlichen Grundlagen', in *Gesammelte Werke*, ed. Birgit Recki (Hamburg: Meiner, 1998); Edmund Husserl, *Ideas Pertaining to a Pure Phenomenology and a Phenomenological Philosophy*, trans. R. Rojcewicz and A. Schuwer (Dordrecht: Kluwer, 1989).

14 While the distinction between esoteric and exoteric is Leibniz's own, Russell took it further, pronouncing a division between Leibniz's 'private' and 'public' texts comprising two separate philosophies, only the former of significance. See Russell, *A Critical Exposition of the Philosophy of Leibniz.*

15 For two novel yet different approaches see Mercer's study, which suggests that Leibniz's metaphysics relied heavily on Platonic sources: Mercer, *Leibniz's Metaphysics*. In his recent study Smith highlights the significance of Leibniz's research in biology and life science for the development of his metaphysics; see Justin E. H. Smith, *Divine Machines: Leibniz and the Sciences of Life* (Princeton, New Jersey: Princeton University Press, 2011).

16 For example by Althusser in 'The Underground Current of the Materialism of the Encounter', Louis Althusser, François Matheron, and Olivier Corpet, *Philosophy of the Encounter: Later Writings, 1978–87* (London; New York: Verso, 2006), and later by Antonio Negri in *The Savage Anomaly: The Power of Spinoza's Metaphysics and Politics* (Minneapolis: University of Minnesota Press, 1991).

17 See Benjamin's Dissertation on the Romantics, where he writes that unlimited reflection results in 'formless thinking', *SW*1, 129.

18 Leibniz, 'On Freedom', *Philosophical Papers and Letters*, 264.

19 Benjamin, *SW*1, 168/*GS*1, 92.

20 See Benjamin, *Origin*, 48–50.

21 Kant, 'Über eine Entdeckung, nach der alle neue Kritik der reinen Vernunft durch eine altere entbehrlich gemacht werden soll', quoted in *Monadology and Other Philosophical Writings*, 209.

22 Antognazza, *Leibniz: An Intellectual Biography*, 4.

23 As he writes, '… toute une machinique de la pensée seront ainsi mobilisées pour *faire la multiple* (car « il faut une méthode qui le fasse effectivement ») et répondre enfin à la question « qu'est-ce que la philosophie? » (« la philosophie est la théorie de multiplicités ») quand viens la vieillesse, et l'heure de parler concrètement, au point singulier ou le concept at la création se rapportent l'un d'autre dans la grand identité. Expressionisme=Constructivisme'. Éric Alliez, *Deleuze, Philosophie Virtuelle*, 1 ed., Collection Les Empêcheurs de Penser en Rond (Le Plessis-Robinson: Synthélabo, 1996), 49.

24 Maria Rosa Antognazza's influential biography, for example, highlights Leibniz as man of reconciliation, arguing that there is an 'all embracing synthetic project' underlying his work. See *Leibniz: An Intellectual Biography*, 8.

25 Weber discusses the debt Derrida's concept of 'deconstruction' owes to concepts of 'destruction' formulated by Luther and Heidegger; as Weber notes, in the process of deconstruction 'propositions and artefacts that claim to constitute a comprehensive system are reinscribed with respect to their enabling limits'. See *Benjamin's-abilities*, 282.

26 Deleuze, *Negotiations*, 143.
27 Benjamin, 'Surrealism', *SW*2, 217.
28 Jay, Martin. *Songs of Experience: Modern American and European Variations on a Universal Theme*. Berkeley (Calif.): University of California Press, 2005, 317.
29 Kevin McLaughlin, 'Biophilology: Walter Benjamin's Literary Critical Legacy', *MLN* 133, no. 3 (April 2018): 562–84.
30 In a letter from 1935, for example, Adorno criticized Benjamin's concept of the dialectical image in his first Exposé to the *Arcades Project*. Adorno targets the 'immanent' version of the concept, which, as he writes, 'implies three things: the conception of the dialectical image as belonging to the content of consciousness, even if a collective consciousness; its linear – I would almost say, historical-developmental – relation to the future as utopia; and the notion of the "epoch" as precisely, the self-contained subject corresponding to this particular content of consciousness'. See 'Exchange with Theodor W. Adorno', in *SW*3, 54. Adorno famously rejected Benjamin's essay 'The Paris of the Second Empire in Baudelaire' from publication by the Institute of Social Research, on the grounds that it was located at the 'crossroads of magic and positivism'. Quoted in Eiland, Howard, and Michael William Jennings, *Walter Benjamin: A Critical Life* (Cambridge, MA: The Belknap Press of Harvard University Press, 2014), 623.
31 Deleuze, *Cinema 2*, 264.
32 Benjamin, 'The Work of Art in the Age of Its Technological Reproducibility (Second Version)', *SW 3*, 122.
33 Deleuze, *Cinema 2*, 263.
34 Johannßen, 'Miniaturization: Reading Benjamin in the Digital Age', 645.
35 Ibid, 643.
36 Deleuze and Guattari, 'Conclusion: from Chaos to Brain', *What Is Philosophy?*, 201–18.
37 Deleuze, *The Fold*, 137.

Bibliography

Abel, Günter. *Nietzsche: Die Dynamik Der Willen Zur Macht Und Die Ewige Wiederkehr.* Berin: De Gruyter, 1984.

Adorno, Theodor W. *Kant's Critique of Pure Reason.* Stanford, CA; Cambridge: Stanford University Press; Cambridge University Press, 2001.

Adorno, Theodor W. *Prisms.* Cambridge, MA: MIT Press, 1981.

Adorno, Theodor W. 'The Essay as Form', trans. Robert Hullot-Kentor and Frederic Will, *New German Critique* 32 (Spring–Summer 1984): 151–71.

Adorno, Theodor W, Walter Benjamín, Henri Lonitz and Nicholas Walker. *The Complete Correspondence, 1928–1940.* Cambridge: Polity Press, 1999.

Agamben, Giorgio. *Sovereign Power and Bare Life.* Homo Sacer 1. Stanford, CA: Stanford University Press, 1998.

Agamben, Giorgio. *The Coming Community*, translated by Michael Hardt. Minnesota: University of Minnesota Press, 1993.

Alberro, Alexander, and Sabeth Buchmann, eds. *Art after Conceptual Art.* Generali Foundation Collection Series. Cambridge, MA: MIT Press, 2006.

Alliez, Eric. *Deleuze, Philosophie Virtuelle.* 1st ed. Collection Les Empêcheurs de Penser En Rond. Le Plessis-Robinson: Synthélabo, 1996.

Althusser, Louis, François Matheron and Olivier Corpet. *Philosophy of the Encounter: Later Writings, 1978–87.* London; New York: Verso, 2006.

Alvizu, Josh. 'Montage Monadology: Leibniz, Benjamin, Modernity' presentation at *Benjamin and Leibniz: On Expression*, a conference at Goldsmith University of London, 27–8 June, 2017.

Andersen, Kirsti. *The Geometry of an Art: The History of the Mathematical Theory of Perspective from Alberti to Monge.* Sources and Studies in the History of Mathematics and Physical Sciences. New York; London: Springer, 2007.

Antognazza, Maria Rosa. *Leibniz, an Intellectual Biography.* Cambridge: Cambridge University Press, 2011.

Antognazza, Maria Rosa. 'Leibniz as Historian', in *The Oxford Handbook of Leibniz*, ed. Maria Rosa Antognazza. Oxford; New York: Oxford University Press, 2018, 591–608.

Antognazza, Maria Rosa, ed. *The Oxford Handbook of Leibniz.* Oxford; New York: Oxford University Press, 2018.

Aragon, and Alyson Waters. *Treatise on Style =: Traité Du Style.* French Modernist Library. Lincoln: University of Nebraska Press, 1991.

Aranda, Julieta, Brian Kuan Wood and Anton Vidokle, eds. *The Internet Does Not Exist.* Berlin: Sternberg Press, 2015.

Arendt, Hannah. *The Human Condition*. 2nd ed. Chicago: University of Chicago Press, 1998.
Aristoteles. *The Complete Works of Aristotle: The Revised Oxford Translation. 1* Princeton, NJ: Princeton University Press, 1995.
Arthur, T. W. Richard. 'The Labyrinth of the Continuum', in *The Oxford Handbook of Leibniz*, ed. Maria Rosa Antognazza. Oxford: Oxford University Press, 2018, 275–89.
Badiou, Alain. *Deleuze: la clameur de l'être*. Paris: A. Fayard, 2010.
Badiou, Alain. 'Gilles Deleuze, the Fold: Leibniz and the Baroque', in *Gilles Deleuze and the Theater of Philosophy*, ed. Constantin V. Boundas and Dorothea Olkowski. London: Routledge, 2019, 51–69.
Baker, Lang. 'The Cry of the Identicals: The Problem of Inclusion in Deleuze's Reading of Leibniz', *Philosophy Today* 39, no. 2 (Summer 1995): 198–211.
Balibar, Étienne. 'Structuralism: A Destitution of the Subject?', trans. James Swenson, *Differences: A Journal of Feminist Cultural Studies* 14, no. 1 (2003): 1–21.
Barbisan, Léa. 'Eccentric Bodies, from Phenomenology to Marxism – Walter Benjamin's Reflections on Embodiment', *Anthropology and Materialism*, Special Issue (2017) 'Discontinuous Infinities', http://journals.openedition.org/am/803.
Barbisan, Léa. *Le corps en exil: Walter Benjamin, penser le corps*. Bibliothèque allemande. Paris: Éditions de la Maison des sciences de l'homme, 2020.
Beaulieu, Alain, ed. *Gilles Deleuze and Metaphysics*. Lanham: Lexington Books, 2014.
Becker, Andreas. 'Von der zweiten zur anderen Natur. Ein Motiv in Walter Benjamins Kunstwerk-Aufsatz', in *Mediale Spielräume*, ed. Thomas Barth, Christian Betzer and Jens Eder u.a. (Hg.). Marburg: Schüren, 2005 (Film- und Fernsehwissenschaftliches Kolloquium 17), 177–86, https://doi.org/10.25969/mediarep/14431.
Beiser, Frederick C. *The German Historicist Tradition*. Oxford: Oxford University Press, 2011.
Belaval, Yvon. *Études leibniziennes: de Leibniz à Hegel*. Collection Tel 229. Paris: Gallimard, 1993.
Belaval, Yvon. *Leibniz critique de Descartes*. Reproduit. Collection Tel 28. Paris: Gallimard, 1978.
Benjamin, Andrew E. *Walter Benjamin and History*. London; New York: Continuum, 2005.
Benjamin, Andrew, and Dimitris Vardoulakis. *Sparks Will Fly: Benjamin and Heidegger*. Albany: State University of New York Press, 2015.
Benjamin, Andrew, and Peter Osborne, eds. *Walter Benjamin's Philosophy: Destruction and Experience*. Manchester: Clinamen Press, 2000.
Benjamin, Walter. *Ecrits français*, translated by Jean-Maurice Monnoyer. Collection Folio Essais 418. Paris: Gallimard, 2003.
Benjamin, Walter. *Einbahnstrasse*, edited by Detlev Schöttker and Steffen Haug Werke und Nachlass, Band 8. Frankfurt am Main: Suhrkamp, 2009.
Benjamin, Walter. *Gesammelte Schriften*, edited by Rolf Tiedemann and Hermann Schweppenhauser. Frankfurt am Main: Suhrkamp Verlag, 1972–1989.

Benjamin, Walter. 'Rigorous Study of Art', trans. Thomas Y. Levin, *October* 47 (Winter 1988): 84–90.

Benjamin, Walter. *Selected Writings, Volume 1: 1913–1926*, edited by Marcus Bullock and Michael W. Jennings. Cambridge; London: Belknap Press, 2004.

Benjamin, Walter. *Selected Writings, Volume 2: Part 1: 1927–1930*, edited by Michael W. Jennings, Howard Eiland and Gary Smith. Cambridge; London: Belknap Press, 2005.

Benjamin, Walter. *Selected Writings, Volume 2: Part 2: 1931–1934*, edited by Michael W. Jennings, Howard Eiland and Gary Smith. Cambridge; London: Belknap Press, 2005.

Benjamin, Walter. *Selected Writings, Volume 3: 1935–1938*, edited by Howard Eiland and Michael W. Jennings. Cambridge; London: Belknap Press, 2006.

Benjamin, Walter. *Selected Writings, Volume 4: 1938–1940*, edited by Howard Eiland and Michael W. Jennings. Cambridge; London: Belknap Press, 2006.

Benjamin, Walter. *The Arcades Project*, edited by Rolf Tiedemann, translated by Howard Eiland and Kevin McLaughlin. Cambridge, MA: Belknap Press, 1999.

Benjamin, Walter. *The Origin of German Tragic Drama*. London; New York: Verso, 2003.

Benjamin, Walter. *Über den Begriff der Geschichte*, edited by Gérard Raulet, 1. Aufl, Werke Und Nachlaß/Walter Benjamin, Bd. 19. Berlin: Suhrkamp, 2010.

Benjamin, Walter. *Understanding Brecht*. London: Verso, 2003.

Benjamin, Walter, Christoph Gödde and Henri Lonitz. *Gesammelte Briefe*. 1. Aufl. Frankfurt am Main: Suhrkamp, 1995.

Benjamin, Walter, Gershom Scholem and Theodor W. Adorno. *The Correspondence of Walter Benjamin, 1910–1940*. Chicago: University of Chicago Press, 1994.

Berg, Olaf. 'Benjamin und Deleuze: Ansätze für eine kritische Geschichtswissenschaft in Filmbildern', *Zeitschrift für kritische Theorie* 12 (2006): 68–97.

Bergman. 'Salomon Maimon', in *A History of Philosophy, Systems of Post-Kantian Philosophy* [תולדות הפילוסופיה החדשה: שיטות בפילוסופיה שלאחר קאנט], ed. Samuel Hugo. Bialik: Jerusalem, 1979, 27–14.

Bergson, Henri. *Matter and Memory*. New York: Zone Books, 1988.

Blanqui, Auguste. *Eternity by the Stars: An Astronomical Hypothesis*, translated by Frank Chouraqui, First Contra Mundum Press edition. New York: Contra Mundum Press, 2013.

Blumenfeld, David. 'Leibniz on Contingency and Infinite Analysis', *Philosophy and Phenomenological Research* XLV, no. 4 (June 1985): 483–514.

Boethius, Anicius Manlius Severinus. '*Consolation of Philosophy*', translated by Joel C. Relihan. Hackett: Indianapolis, 2001.

Boundas, Constantin V., and Olkowski Dorothea, eds. *Gilles Deleuze and the Theater of Philosophy*. New York: Routledge, 1994.

Breton, André. *Manifestes du Surréalisme*. Paris: Gallimard, 1966, 11–64.

Breton, André. *Manifestoes of Surrealism*. 1. ed. as an Ann Arbor paperback. Ann Arbor Paperbacks 182. Ann Arbor, MI: University of Michigan Press, 1972.

Breton, André. *Nadja*. New York: Grove Press, 1999.

Brown, Richard C. *The Tangled Origins of the Leibnizian Calculus: A Case Study of a Mathematical Revolution*. Singapore; Hackensack, NJ: World Scientific, 2012.

Buchanan, Ian, ed. *A Deleuzian Century?*. Durham, NC: Duke University Press, 1999.

Buci-Glucksmann, Christine. 'Les cristaux de l'art: une esthétique du virtuel', *Rue Descartes* 20 (May 1998): 95–111.

Buck-Morss, Susan. 'Aesthetics and Anaesthetics: Walter Benjamin's Artwork Essay Reconsidered', *October* 62 (Autumn 1992): 3–41.

Buck-Morss, Susan. *The Dialectics of Seeing: Walter Benjamin and the Arcades Project*. 8. print. Cambridge, MA: MIT Press, 1999.

Bürger, Peter. *Theory of the Avant-Garde*. Theory and History of Literature 4. Manchester: Manchester University Press [u.a.], 1984.

Cache, Bernard, *Earth Moves: The Furnishing of Territories*, translated by Anne Boyman and Michael Speaks. 10. [pr.]. Writing Architecture. Cambridge, MA: MIT Press, 2010.

Cache, Bernard, and Christian Girard. 'Objectile: The Pursuit of Philosophy by Other Means?', in *Deleuze and Architecture*, Deleuze Connections, ed. Hélène Frichot. Edinburgh: Edinburgh University Press, 2013, 96–110.

Cadava, Eduardo. *Words of Light: Theses on the Photography of History*. 2. print., 1. paperback print. Princeton Paperbacks. Princeton, NJ: Princeton University Press, 1998.

Cameron, Margaret. 'Ac pene Stoicus: Valla and Leibniz on "The Consolation of Philosophy"', *History of Philosophy Quarterly* 24, no. 4 (October 2007): 337–54.

Campbell, Iain. 'Structuralist heroes and points of heresy: Recognizing Gilles Deleuze's (anti-)structuralism', *Cont Philos Rev* 55 (2022): 215–34. https://doi.org/10.1007/s11007-021-09562-9.

Carpo, Mario. *Beyond Digital* and 'A Short but Believable History of the Digital turn in Architecture', https://www.e-flux.com/architecture/chronograms/528659/a-short-but-believable-history-of-the-digital-turn-in-architecture.

Carpo, Mario. 'Introduction', in *Projectiles*, ed. Bernard Cache. Architecture Words 6. London: Architectural Association, 2011, 5–11.

Cassirer, Ernst. *Die Begriffsform Im Mythischen Denken*. Braunschweig: Vieweg and Teubner Verlag, 1922.

Cassirer, Ernst, and Birgit Recki. *Gesammelte Werke*. Hamburger Ausg. Hamburg: Meiner, 1998.

Caygill, Howard. *Walter Benjamin the Colour of Experience*. London; New York: Routledge, 1998.

Chakrabarty, Dipesh. 'Decentring the Human?', *The Human Condition in the Anthropocene*, The Tanner Lectures in Human Values, Delivered at Yale University, 18–19 February 2015.

Chakrabarty, Dipesh. *The Climate of History in a Planetary Age*. Chicago; London: The University of Chicago Press, 2021.

Cohen, Hermann. *Das Prinzip der Infinitesimal-Methode und seine Geschichte, ein Kapitel zur Grundlegung der Erkenntniskritik.* Frankfurt am Main: Suhrkamp, 1968.

Cohen, Hermann. *Kants Theorie der Erfahrung.* Saarbrücken: VDM, Müller, 2006.

Cohen, Hermann. *System der Philosophie; Teil 1. Logik der reinen Erkenntnis*, ed. Helmut Holzhey. Hildesheim: Olms, 2005.

Cohen, Margaret. 'Walter Benjamin's Phantasmagoria', *New German Critique* 48 (Autumn 1989): 87–107.

Conley, Tom. 'From Multiplicities to Folds: On Style and Form in Deleuze', in *A Deleuzian Century?*, ed. Ian Buchman. Durham, NC: Duke University Press, 1999, 574–615.

Copeland, Rita, and Peter T. Struck, eds. *The Cambridge Companion to Allegory*, Cambridge Companions to Literature. Cambridge; New York: Cambridge University Press, 2010.

Coudert, Allison. *Leibniz and the Kabbalah.* International Archives of the History of Ideas 142. Dordrect; Boston: Kluwer Academic, 1995.

Couturat, Louis. *La logique de Leibniz d'après des documents inédits*, Paris: Ancienne Librairie Germer Baillière, 1901.

Couturat, Louis. 'Sur la Métaphysique de Leibniz (avec un opuscule inédit)', *Revue de Métaphysique et de Morale*, 10, no. 1 (Janvier 1902): 1–25, Presses Universitaires de France.

Croce, Benedetto. *The Essence of Aesthetic.* Translated by Douglas Ainslie. London: Heineman, 1921.

Culp, Andrew. *Dark Deleuze.* Minneapolis: University of Minnesota Press, 2016.

Damisch, Hubert. *The Origin of Perspective*, 3. Print. Cambridge, MA: MIT Press, 2000.

Davillé, Louis. *Leibniz Historien, Essai sur l'activite et la methode Historiques de Leibniz.* Paris: Librairie Felix Alcan, 1909.

Davis, Martin. *The Universal Computer: The Road from Leibniz to Turing.* New York: Norton, 2000.

De Buzon, Frédéric. 'Que lire dans le deux infinis ? Remarques sur une lecture Leibnizienne', Presses Universitaires de France, *Les Études philosophiques* 4 (2010): 95, 535–48.

Debuiche, Valerie. 'L'expression leibnizienne et ses modèles mathématiques', *Journal of the History of Philosophy* 51, no. 3 (2013): 409–39.

Debuiche, Valerie. *Leibniz: un philosophe savant.* Aimer les philosophes. Paris: Ellipses, 2017.

DeLanda, Manuel. 'Immanence and Transcendence in the Genesis of Form', in *A Deleuzian Century?*, ed. Ian Buchman. Durham, NC: Duke University Press, 1999, 281–317.

Deleuze, Gilles. *Bergsonism.* New York: Zone Books, 1988.

Deleuze, Gilles. *Cinema 1.* London; New York: Bloomsbury, 1986.

Deleuze, Gilles. *Cinema 2.* London; New York: Bloomsbury, 2013.

Deleuze, Gilles. *Desert Islands and Other Texts, 1953–1974.* Semiotext(e) Foreign Agents Series. Los Angeles, CA; Cambridge, MA: Semiotext(e); Distributed by MIT Press, 2004.

Deleuze, Gilles. *Difference and Repetition.* translated by Paul Patton. London: Continuum, 2004.

Deleuze, Gilles. *Différence et répétition*. 4. éd. Bibliothèque de philosophie contemporaine. Paris: Presses Universitaires de France, 1981.

Deleuze, Gilles. *Essays Critical and Clinical*, translated by Daniel W. Smith and Michael A. Greco. London; New York: Verso, 1998.

Deleuze, Gilles. *Expressionism in Philosophy: Spinoza*. New York; Cambridge, MA: Zone Books; Distributed by MIT Press, 1990.

Deleuze, Gilles. *Kant's Critical Philosophy: The Doctrine of the Faculties*. London: Athlone Press, 1984.

Deleuze, Gilles. *L'image-Movement*. Paris: Editions de Minuit, 1998.

Deleuze, Gilles. *L'image-Temps*. Collection 'Critique' 2. Paris: Editions de Minuit, 1985.

Deleuze, Gilles. *La philosophie critique de Kant*. Paris: Presses Universitaires de France, 2015.

Deleuze, Gilles. *Le pli: Leibniz et le Baroque*. Nachdr. Collection critique. Paris: Éd. de Minuit, 2005.

Deleuze, Gilles. *Mille Plateaux*. Collection 'Critique', t. 2. Paris: Éditions de minuit, 1980.

Deleuze, Gilles. *Negotiations: 1972–1990*. New York: Columbia University Press, 1995.

Deleuze, Gilles. *Nietzsche and Philosophy*. European Perspectives. New York: Columbia University Press, 2006.

Deleuze, Gilles. *Nietzsche et la philosophie*. 7e édition 'Quadrige'. Quadrige. Paris: Presses Universitaires de France, 2014.

Deleuze, Gilles. *Proust and Signs*, translated by Richard Howard. London; New York: Continuum, 2008.

Deleuze, Gilles. *Spinoza et le problème de l'expression*. Paris: Editions de Minuit, 1969.

Deleuze, Gilles. *Cours Vincennes*, lectures given at Paris 8-Vincennes – Saint-Denis University, 'Sur Kant' (1978); 'Sur Leibniz' (1980); 'Cours sur le cinéma' (1981–82); 'Sur le cinéma: L'image mouvement et l'image temps' (1982); 'Sur le cinéma: Classifications des signes et du temps' (1982–83); 'Sur le cinéma: l'image pensée' (1984–85); 'Sur Leibniz: Leibniz et le baroque' (1986–87); 'Sur Leibniz: Les principes et la liberté' (1987).

Deleuze, Gilles. 'The Exausted', trans. Anthony Uhlmann, *Substance* 24, No. 3, no. 78 (1995): 3–28.

Deleuze, Gilles. *The Fold: Leibniz and the Baroque*. London: The Athlone Press, 1993.

Deleuze, Gilles. *The Logic of Sense*, translated by Mark Lester, Charles J Stivale and Constantin V Boundas. New York: Columbia University Press, 1990.

Deleuze, Gilles, and Claire Parnet. 'E as in Enfance', in *Gilles Deleuze from A to Z*, Semiotext(e) Foreign Agents Series. Cambridge, MA: Distributed by The MIT Press, 2011.

Deleuze, Gilles, and Felix Guattari. *A Thousand Plateaus: Capitalism and Schizophrenia*. Minneapolis: University of Minnesota Press, 1987.

Deleuze, Gilles, and Felix Guattari. *Anti-Oedipus: Capitalism and Schizophrenia.* Minneapolis: University of Minnesota Press, 1983.

Deleuze, Gilles, and Felix Guattari. *Kafka: Toward a Minor Literature.* 9. print. Theory and History of Literature 30. Minneapolis: University of Minnesota Press, 2008.

Deleuze, Gilles, and Felix Guattari. *L' anti-OEdipe.* Capitalisme et schizophrenie, Gilles Deleuze; Felix Guattari; 1. Paris: Ed. de Minuit, 2012.

Deleuze, Gilles, and Felix Guattari. *Qu'est-ce que la philosophie?* Nachdr. Collection Critique. Paris: Ed. de Minuit, 1991.

Deleuze, Gilles, and Felix Guattari. *What Is Philosophy?* Repr. London: Verso, 2011.

Denson, Shane. *Discorrelated Images.* Durham: Duke University Press, 2020.

Denson, Shane, and Julia Leyda, eds. *Post-Cinema: Theorizing 21st-Century Film.* Falmer: REFRAME Books, 2016.

De Risi, Vincenzo. 'Analysis Situs, the Foundation of Mathematics, and a Geometry of Space', in *The Oxford Handbook of Leibniz*, ed. Maria R. Antognazza. Oxford: Oxford University Press, 2018, 247–58.

Desargues, Girard. *Brouillon project d'une atteinte aux événements des rencontres du cône avec un plan.* In *L'œuvre mathématique de G. Desargues*, 2nd ed., edited by René Taton. Paris: Vrin, 1981, 99–180.

Descartes, René. *Discourse on Method and Meditations*, translated by Elizabeth Sanderson Haldane and George Robert Thomson Ross, Dover Philosophical Classics. Mineola, NY: Dover Publications, 2003.

Descartes, René. *Méditations métaphysiques.* 3. éd. Les grands textes, edited by Florence Khodoss. Paris: Presses Universitaires de France, 1992.

Descartes, René. *Principles of Philosophy.* Synthese Historical Library: Texts and Studies in the History of Logic and Philosophy, translated by Valentine Rodger Miller and Reese P. Miller. v. 24. Dordrecht; Boston; Hingham, MA: Reidel; Distributed by Kluwer Boston, 1983.

Descombes, Vincent. *Modern French Philosophy.* Cambridge; New York: Cambridge University Press, 1980.

Desideri, Fabrizio. 'Intermittency: The Differential of Time and the Integral of Space, the Intensive Spatiality of the Monad, the *Apokatastasis* and the Messianic World in Benjamin's Latest Thinking', *Aisthesis rivista on-line del Seminario Permanente di Estetica* IX, no. 1 (2016): 177–87.

De Wilde, Marc. 'Meeting Opposites: The Political Theologies of Walter Benjamin and Carl Schmitt', *Philosophy & Rhetoric* 44, no. 4 (2011): 363–81.

Di Bella, Stefano. *The Science of the Individual: Leibniz's Ontology of Individual Substance.* Dordrecht: Springer Netherlands, 2005.

Döblin, Alfred. *Berlin Alexanderplatz*, translated by Michael Hofmann. NYRB Classics. New York: New York Review Books, 2018.

Dosse, François. *Gilles Deleuze et Félix Guattari: biographie croisée.* Paris: La Découverte, 2009.

Duffy, Simon B. *Deleuze and the History of Mathematics: In Defence of the 'New'.* Bloomsbury Studies in Continental Philosophy. London; New York: Bloomsbury, 2013.

Dvořák, Max. 'Über Greco und den Manierismus', in *Kunstgeschichte als Geistesgeschichte. Studien zur abendländischen Kunstentwicklung*. Munich: Piper, 1924, 259–76.

Eagleton, Terry. *The Ideology of the Aesthetic*. Oxford; Cambridge, MA: Blackwell, 1990.

Eiland, Howard, and Michael William Jennings. *Walter Benjamin: A Critical Life*. Cambridge, MA: The Belknap Press of Harvard University Press, 2014.

Fenves, Peter D. *Arresting Language: From Leibniz to Benjamin*. Meridian. Stanford, CA: Stanford University Press, 2001.

Fenves, Peter D. *The Messianic Reduction: Walter Benjamin and the Shape of Time*. Meridian, Crossing Aesthetics. Stanford, CA: Stanford University Press, 2011.

Ferber, Ilit. *Philosophy and Melancholy: Benjamin's Early Reflections on Theater and Language*. Cultural Memory in the Present. Stanford, CA: Stanford University Press, 2013.

Feuerbach, Ludwig. 'Towards a Critique of Hegel's Philosophy' (1839), in *The Fiery Brook: Selected Writings*, trans. Zawar Hanfi. London: Verso, 2013, 53–98.

Flanagan, Timothy. *Baroque Naturalism in Benjamin and Deleuze: The Art of Least Distances*. Cham: Palgrave Macmillan, 2021.

Flanagan, Timothy. 'Presentation of Reciprocity: An Interpolative Study of the Baroque Works of Walter Benjamin and Gilles Deleuze', PhD diss., University of Dundee, 2007.

Flanagan, Timothy. 'The Free and Indeterminate Accord of "The New Harmony": The Significance of Benjamin's Study of the Baroque for Deleuze', in *Deleuze and The Fold: A Critical Reader*, ed. Sjoerd Tuinen and Niamh McDonnell. Basingstoke; New York: Palgrave Macmillan, 2010, 46–64.

Frichot, Hélène. 'Deleuze and the Story of the Superfold', in *Deleuze and Architecture*, Deleuze Connections, ed. Hélène Frichot. Edinburgh: Edinburgh University Press, 2013, 79–95.

Friedlaender, Walter. *Mannerism and Anti-Mannerism in Italian Painting*, 9. Print. New York: Schocken Books, 1985, 8.

Friedlander, Eli. *Walter Benjamin: A Philosophical Portrait*. Cambridge, MA: Harvard University Press, 2012.

Futch, Michael. *Leibniz's Metaphysics of Time and Space*. New York: Springer, 2014.

Gal, Ofer, and Raz Chen-Morris, eds. *Science in the Age of Baroque*. Archives Internationales d'histoire Des Idées = International Archives of the History of Ideas 208. Dordrecht; New York: Springer, 2013.

Gentry, George. 'The Subject in Whitehead's Philosophy', *Philosophy of Science* 11, no. 4 (October 1944): 222–6.

Giovanelli, Marco. *Reality and Negation – Kant's Principle of Anticipations of Perception: An Investigation of Its Impact on the Post-Kantian Debate*. Studies in German Idealism, volume 11. Dordrecht; Heidelberg; London; New York: Springer, 2011.

Gould, Rebecca. 'Laws, Exceptions, Norms: Kierkegaard, Schmitt, and Benjamin on the Exception', *Telos: Critical Theory of the Contemporary* 162 (2013): 77–96.

Guéroult, Martial. *La Philosophie transcendantale de Salomon Maimon*. Paris: Alcan, 1929.

Hallward, Peter. *Out of This World: Deleuze and the Philosophy of Creation*. London; New York: Verso, 2006.

Hamacher, Werner. 'En-counterings of Time', trans. Peter Hanly. *Philosophy Today* 60, no. 4 (Fall 2016): 839–49.

Hamacher, Werner. 'Intensive Languages', *MLN* 127 (2012): 485–541.

Hamacher, Werner. 'Now: Walter Benjamin and Historical Time', in *Walter Benjamin and History*, ed. Andrew E. Benjamin. London; New York: Continuum, 2005, 38–68.

Hanssen, Beatrice. 'Philosophy at Its Origin: Walter Benjamin's Prologue to the *Ursprung des deutschen Trauerspiels*', *MLN* 110, no. 4 (September 1995): 809–33.

Hart, David Bentley, ed. *The New Testament: A Translation*. New Haven: Yale University Press, 2017.

Heidegger, Martin. *Being and Time*, trans. Joan Stambaugh. SUNY Series in Contemporary Continental Philosophy. Albany: State University of New York Press, 2010.

Heil, Susanne. '*Gefährliche Beziehungen*': *Walter Benjamin und Carl Schmitt*. Stuttgart: Metzler, 1996.

Herbertz, Richard. *Die Lehre vom Unbewussten im System des Leibniz*. Niemeyer: Halle, 1905.

Hills, Helen. 'The Grit in the Oyster of Art History', in *Rethinking the Baroque*, ed. Helen Hills. Surrey; Burlington, VT: ASHGATE, 2011, 11–38.

Hobbes, Thomas. 'Leviathan', *Cambridge Texts in the History of Political Thought*, ed. Richard Tuck. Cambridge; New York: Cambridge University Press, 1996.

Hofmann, Hasso. 'Souverän ist, wer uber die Ausnahmezustand Entscheidet', *Der Staat* 44, no. 2 (2005): 171–86.

Holdsworth, David. 'Philosophical Problematisation and Mathematical Solution: Learning Science with Gilles Deleuze', *Deleuze and Education*, Deleuze Connections, ed. Inna R. Semetsky and Diana Masny. Edinburgh: Edinburgh University Press, 2013.

Holzhey, Helmut. *Natorp und Cohen*. Basel: Schwabe, 1986, 137–54.

Horkheimer, Max. 'On Bergson's Metaphysics of Time', *Radical Philosophy* 131 (May–June 2005), https://www.radicalphilosophy.com/article/on-bergsons-metaphysics-of-time.

Hosman, Michael. 'Benjamin's Baroque', *Thresholds* 2005, no. 28 (2005): 119–29.

Husserl, Edmund. *Ideas Pertaining to a Pure Phenomenology and to a Phenomenological Philosophy*, First Book: General Introduction to a Pure Phenomenology, translated by F. Kersten. The Hague: Martinus Nijhoff Publishers, 1983.

Jay, Martin. *Downcast Eyes*. California: University of California Press, 1993.

Jay, Martin. *Songs of Experience: Modern American and European Variations on a Universal Theme*. Berkeley, CA: University of California Press, 2005.

Jay, Martin. *The Dialectical Imagination: A History of the Frankfurt School and the Institute of Social Research, 1923–1950*. Weimar and Now 10. Berkeley: University of California Press, 1996.

Johannßen, Dennis. 'Gottfried Wilhelm Leibniz. Andacht zum Kleinen: zu Benjamins Leibniz-Lektüren', in *Entwendungen: Walter Benjamin und seine Quellen*, ed. Nadine Werner and Jessica Nietzsche. Paderborn: Wilhelm Fink, 2018, 281–302.

Johannßen, Dennis. 'Miniaturization: Reading Benjamin in the Digital Age', *MLN* 133, no. 3 (April 2018): 637–53.

Jones, Graham, ed. *Deleuze's Philosophical Lineage*. Repr. Edinburgh: Edinburgh University Press, 2009.

Jones, Graham. 'Solomon Maimon', *Deleuze's Philosophical Lineage*, 104–29.

Kant, Immanuel. *Critique of the Power of Judgment*, ed. Paul Guyer, trans. Eric Matthews. The Cambridge Edition of the Works of Immanuel Kant. Cambridge: Cambridge University Press, 2013.

Kant, Immanuel. 'Idea for a Universal History from a Cosmopolitan Perspective', in *Towards Perpetual Peace and Other Writings on Politics, Peace and History*, ed. Pauline Kleingeld, trans. David L. Colclasure. New Haven: Yale University Press, 2006, 3–16.

Kant, Immanuel. 'On the Basis of the Difference of Regions in Space' (1768), in *Gesammelte Schriften*, ed. Akademie der Wissenschaften. Berlin: Reimer, de Gruyter, 1900, vol. 2, 378.

Kant, Immanuel. *Prolegomena to Any Future Metaphysics That Will Be Able to Come Forward as Science: With Selections from the Critique of Pure Reason*. Cambridge Texts in the History of Philosophy, edited by Gary C. Hatfield. Cambridge; New York: Cambridge University Press, 2004.

Kant, Immanuel, and Mary J. Gregor. *Critique of Practical Reason*. 13. print. Cambridge Texts in the History of Philosophy. Cambridge: Cambridge University Press, 2012.

Kant, Immanuel, Paul Guyer and Allen W. Wood. *Critique of Pure Reason*. The Cambridge Edition of the Works of Immanuel Kant. Cambridge; New York: Cambridge University Press, 1998.

Kepler, Johannes. *The Harmony of the World*, edited by Eric J. Aiton, Memoirs of the American Philosophical Society 209. Philadelphia, PA: American Philosophical Society, 1997.

Kerslake, Christian. *Immanence and the Vertigo of Philosophy: From Kant to Deleuze*. Plateaus: New Directions in Deleuze Studies. Edinburgh: Edinburgh University Press, 2009.

Kholeif, Omar ed., *Art in the Age of Anxiety* (Cambridge: The MIT Press, 2020).

Klenner, Hermann. 'Savigny's Research Program of the Historical School of Law and Its Intellectual Impact in 19th Century Berlin', *The American Journal of Comparative Law* 37, no. 1 (Winter 1989): 69–70.

Koselleck, Reinhart. *Futures Past: On the Semantics of Historical Time*. New York: Columbia University Press, 2004.

Kuleshov, Lev. 'Kuleshov Effect', 1921, https://www.youtube.com/watch?v=_gGl3LJ7vHc.

Lambert, Gregg, and Gregg Lambert. *On the (New) Baroque*. Rev. ed. Critical Studies in the Humanities. Aurora, CO: Davies Group, 2008.

Leibniz, Gottfried Wilhelm. 'Apokatastasis Panton' (1715) translated by Forman David from M. Fichant (ed.), De l' Horizon de la Doctrine Humaine (1693); Ἀποκατάστασις πάντων (La Restitution universelle) (1715). Paris: J. Vrin, 1991, 60–6. https://philarchive.org/archive/LEIAPA-4.

Leibniz, Gottfried Wilhelm. *De l'horizon de la doctrine humaine: 1693*, edited by Michel Fichant, Bibliothèque des textes philosophiques. Paris: Librairie philosophique, 1991.

Leibniz, Gottfried Wilhelm. 'Essai anagogique dans la recherche des causes *Tentamen anagogicum*' (1695), in *Die philosophischen Schriften*, ed. Karl Immanuel Gerhardt. Berlin: Weidmann, 1890, bd.7, 270–9.

Leibniz, Gottfried Wilhelm. *La Monadologie: précédée d'une étude de Jaques Rivelaygue*, edited by Émile Boutroux. Le livre de poche classique de la philosophie 4606. Paris: Librairie Générale Française, 1991.

Leibniz, Gottfried Wilhelm. *La naissance du calcul différentiel*, edited by Marc Parmentier and Michel Serres, Mathesis. Paris: J. Vrin, 1989.

Leibniz, Gottfried Wilhelm. *Leibniz's Monadology: A New Translation and Guide*, edited by Lloyd Strickland. Edinburgh: Edinburgh University Press, 2014.

Leibniz, Gottfried Wilhelm. *New Essays on Human Understanding*, edited by Peter Remnant and Jonathan Bennett. Cambridge Texts in the History of Philosophy. Cambridge; New York: Cambridge University Press, 1996.

Leibniz, Gottfried Wilhelm. *Philosophical Papers and Letters*, edited by Leroy E. Loemker. 2.ed., 2. print. Synthese Historical Library, v. 2. Dordrecht; Boston: D. Reidel Pub. Co, 1976.

Leibniz, Gottfried Wilhelm. *Theodicy: Essays on the Goodness of God, the Freedom of Man, and the Origin of Evil*, translated by E. M. Huggard and Austin Farrer. New York: Cosimo, 2009.

Leibniz, Gottfried Wilhelm. *The Monadology and Other Philosophical Writings*, edited by Robert Latta. London: Oxford University Press, 1971.

Leibniz, Gottfried Wilhelm, Roger Ariew and Daniel Garber. *Philosophical Essays*. Indianapolis: Hackett Pub. Co, 1989.

Leinkauf, Thomas. 'Leibniz und sein Geschichtsbegriff', *Studia Leibnitiana*, 51 (2019): 12–41.

Lenain, Thierry. *L' image: Deleuze, Foucault, Lyotard*. Paris: J. Vrin, 1997.

Levin, Noa. 'Montage Mahagonny: Walter Benjamin and Bertolt Brecht's Theatre of Interruptions', in *Material und Begriff. Arbeitsverfahren und theoretische Beziehungen Walter Benjamins*. Hamburg: Argument, 2019, 145–59.

Levin, Noa. 'Specters of Eternal Return: Benjamin and Deleuze Read Leibniz', *Filozofski vestnik*, special issue on 'world' XLII, no. 2 (2021): 305–29.

Levin, Thomas Y. 'Walter Benjamin and the Theory of Art History', *October* 47 (Winter 1988): 77–83.

Lévy, Pierre. *Becoming Virtual: Reality in the Digital Age*. New York: Plenum Trade, 1998.

Lewis, David K. *On the Plurality of Worlds*. Oxford: Blackwell, 1986.

Locke, John. *An Essay Concerning Human Understanding*, trans. Roger Stuart Woolhouse. London; New York : Penguin Books, 1997.

Loemeker, Leroy E. 'Introduction: Leibniz as Philosopher', in *Philosophical Papers and Letters*, 2.ed., 2. print. Synthese Historical Library, v. 2. Dordrecht; Boston: D. Reidel Pub. Co, 1976.

Lohenstein, Daniel Casper von. *Sophonisbe: Trauerspiel*, edited by Rolf Tarot. Bibliographisch aktualisierte Ausgabe. Reclams Universal-Bibliothek, Nr. 14234. Ditzingen: Reclam, 2021.

Lotze, Hermann. *Mikrokosmus*. Leipzig: Hirzel, 1896–1909.

Löwy, Michael. *Fire Alarm: Reading Walter Benjamin's 'On the Concept of History'*. London; New York: Verso, 2016.

Lukács, Georg. *The Theory of the Novel: A Historico-Philosophical Essay on the Forms of Great Epic Literature*, 12. Print. Cambridge, MA: The MIT Press, 1994.

Lundy, Craig. 'From Structuralism to Poststructuralism', in *The Edinburgh Companion to Poststructuralism*, ed. Benoît Dillet, Iain Mackenzie and Robert Porter. Edinburgh: Edinburgh University Press, 2013, 69–92.

Maillet, Arnaud. *The Claude Glass: Use and Meaning of the Black Mirror in Western Art*, First paperback edition. New York: Zone Books, 2009.

Maimon, Salomon. *Essay on Transcendental Philosophy*. London; New York: Continuum, 2010.

Maimon, Salomon. *Versuch über die Transzendentalphilosophie*, edited by Florian Ehrensperger Philosophische Bibliothek 552. Hamburg: Meiner, 2004.

Mallarmé, Stéphane. *Un coup de dés jamais n'abolira le hasard: poème*. Faksimile-Edition der Ausgabe 1914. Paris: Gallimard, 2014.

Mallarmé, Stéphane, and Bertrand Marchal. *Œuvres Complètes*. Nouv. éd. Bibliothèque de La Pléiade 65, 497. Paris: Gallimard, 1998.

Mallarmé, Stéphane, and Henry Weinfield. *Collected Poems: A Bilingual Edition*. Berkeley: University of California Press, 1996.

Manzo, Fiorenza. 'How Sincere Was Leibniz's Criticism of Hobbes' Political Thought?', *The Leibniz Review* 30, 2020, https://doi.org/10.5840/leibniz2020302.

Massimi, Michela. *Perspectival Realism*, Oxford Studies in Philos Science Series. New York: Oxford University Press, 2022.

Massimi, Michela, and D. McCoy Casey, eds. *Understanding Perspectivism*. London: Routledge, 2019.

McBride, Patrizia C. *The Chatter of the Visible: Montage and Narrative in Weimar Germany*. Ann Arbor: University of Michigan Press, 2016.

McDonough, Jeffrey K. 'Optics', in *The Oxford Handbook of Leibniz*, ed. Maria Rosa Antognazza. Oxford: Oxford University Press, 2018, 425–37.

Mclaughlin, Kevin. 'Biophilology: Walter Benjamin's Literary Critical Legacy', *MLN* 133, no. 3 (April 2018): 562–84.

Mclaughlin, Kevin. 'Virtual Paris: Benjamin's Arcades Project', in *Benjamin's Ghosts: Interventions in Contemporary Literary and Cultural Theory*, ed. Gerhard Richter. Stanford, CA: Stanford University Press, 2002, 204–25.

Meillassoux, Quentin, and Robin Mackay. *The Number and the Siren: A Decipherment of Mallarmé's Coup de Dés*. Falmouth; New York: Urbanomic: Sequence Press, 2012.

Meinecke, Friedrich. *Die Entstehung des Historismus*, ed. Carl Hinrichs and Hans Herzfeld. 3rd ed. Berlin; Boston: Walter de Gruyter GmbH, 2016.

Meinecke, Friedrich, Hans Herzfeld and Carl Hinrichs. *Werke. 3, 3.* Stuttgart: Koehler, 1959.

Melchior-Bonnet, Sabine. *The Mirror: A History*, translated by Katharine H. Jewett. London; New York: Routledge, 2014.

Mercer, Christia. *Leibniz's Metaphysics: Its Origins and Development.* Cambridge; New York: Cambridge University Press, 2001.

Mercier, Lucie Kim-Chi. 'Michel Serres's Leibnizian Structuralism', *Angelaki* 24, no. 6 (2019): 3–21.

Mormann, Thomas, and Mikhail Katz. 'Infinitesimals as an Issue of Neo-Kantian Philosophy of Science', *HOPOS: The Journal of the International Society for the History of Philosophy of Science* 3, no. 2 (Fall 2013): 236–80.

Murray, Timothy. *Digital Baroque: New Media Art and Cinematic Folds.* Minneapolis: University of Minnesota Press, 2008.

Nachtomy, Ohad. *Living Mirrors: Infinity, Unity, and Life in Leibniz's Philosophy.* New York: Oxford University Press, 2019.

Negri, Antonio. *The Savage Anomaly: The Power of Spinoza's Metaphysics and Politics.* Minneapolis: University of Minnesota Press, 1991.

Newman, Jane O. *Benjamin's Library: Modernity, Nation, and the Baroque.* Signale : Modern German Letters, Cultures, and Thought. Ithaca, NY: Cornell University Press: Cornell University Library, 2011.

Ng, Julia. 'Acts of Time: Cohen and Benjamin on Mathematics and History', *Paradigmi. Rivista di critica filosofica* 1 (2017): 41–60.

Nietzsche, Friedrich. *Thus Spoke Zarathustra: A Book for Everyone and No One*, translated by Reginald John Hollingdale, Repr. with new chronology, Penguin Classics. London: Penguin Books, 1961.

Nitsche, Jessica, and Nadine Werner, eds. *Entwendungen: Walter Benjamin und seine Quellen.* Paderborn: Wilhelm Fink, 2018.

Osborne, Peter. *Anywhere or Not At All: Philosophy of Contemporary Art.* London: Verso, 2013.

Osborne, Peter. *The Politics of Time: Modernity and Avant-Garde.* London; New York: Verso, 1995.

Osborne, Peter. *The Postconceptual Condition: Critical Essays.* Brooklyn: Verso, 2018.

Osborne, Peter, ed. *Walter Benjamin: Critical Evaluations in Cultural Theory.* London; New York: Routledge, 2005.

Panofsky, Erwin, *The Life and Art of Albrecht Dürer*, translated by Jeffrey Chipps Smith, 1st Princeton classic, ed. A Princeton Classic Edition. Princeton, NJ: Princeton University Press, 2005.

Phemister, Pauline. 'Leibniz and Ecology', *History of Philosophy Quarterly* 18, no. 3 (July 2001): 239–58.

Plato. *Plato's Meno*, ed. G. M. A. Grube. Indianapolis: Hackett Pub. Co, 1976.

Plato. *The Being of the Beautiful: Plato's Theaetetus, Sophist, and Statesman*, edited by Seth Benardete. Chicago, IL; Bristol: University of Chicago Press; University Presses Marketing distributor, 2006.

Plato. *The Republic of Plato*, translated by Allan Bloom. New York: Basic Books, 1986.
Plotinsky, Arkady. 'Chaosmologies: Quantum Field Theory, Chaos and Thought in Deleuze and Guattari's What Is Philosophy?', *Paragraph* 29, no. 2, Deleuze and Science (July 2006): 40–56.
Poma, Andrea. *The Critical Philosophy of Hermann Cohen = La Filosofia Critica Di Hermann Cohen*. Albany: State University of New York Press, 1997.
Posteraro, Tano S. 'Not Just Do as I Do: Knowledge and Learning in the Image of Thought', *Deleuze Studies* 9, no. 4 (2015): 459.
Pudovkin, Vsevolod. *Selected Essays*, edited by Richard Taylor. Chicago: University of Chicago Press, 2006.
Rancière, Jacques. *Mallarmé: La politique de la sirène*, Paris: Hachette, 1996.
Rancière, Jacques. 'The Concept of Anachronism and the Historian's Truth', *InPrint* 3, no. 1, Article 3 (2015): 21–52.
Rateau, Paul. *La question du mal chez Leibniz: fondements et élaboration de la Théodicée*. Travaux de philosophie 15. Paris: Champion, 2008.
Ray, Man, dir. *Le Retour à la raison*. Alice Prin, 1923, 3 min.
Renais, Alain, dir. *Last Year in Marienbad*. Cineriz: France, 1961.
Rescher, Nicholas. *The Philosophy of Leibniz*. New Jersey: Prentice-Hall, 1967.
Rey, Anne-Lise. 'L'ambivalence de la notion d'action dans la Dynamique de Leibniz. La correspondance entre Leibniz et De Volder', *Studia Leibnitiana* 41, no. 1 (2009): 47–66.
Ricardou, Jean. *Le Nouveau Roman*, Points Littérature 212. Paris: Ed. du Seuil, 1990.
Richards, Robert J. *The Romantic Conception of Life: Science and Philosophy in the Age of Goethe*. Chicago: University of Chicago Press, 2002.
Richter, Gerhard, ed. *Benjamin's Ghosts: Interventions in Contemporary Literary and Cultural Theory*. Stanford, CA: Stanford University Press, 2002.
Richter, Hans, dir. 'Ghosts Before Breakfast' (*Vormittagsspuk*). Richter Hans, 1928, 9 min. https://www.youtube.com/watch?v=bc2eDeYRglM.
Ricœur, Paul. *Memory, History, Forgetting*, translated by Kathleen Blamey and David Pellauer. Paperback ed., [Nachdr.]. Chicago, IL: University of Chicago Press, 2010.
Riegl, Alois. *Late Roman Art Industry*, trans. Rolf Winkes, Archaeologica 36. Rome: G. Bretschneider, 1985.
Roberts, John T. 'Leibniz on Force and Absolute Motion', *Philosophy of Science* 70, no. 3 (July 2003): 553–73.
Ross, Alison. *Walter Benjamin's Concept of the Image*. First issued in paperback. Routledge Studies in Twentieth-Century Philosophy 38. London; New York: Routledge, Taylor & Francis Group, 2015.
Roudinesco, Elisabeth. *Philosophy in Turbulent Times: Canguilhem, Sartre, Foucault, Althusser, Deleuze, Derrida*. Paperback ed. Chichester, NY: Columbia University Press, 2010.
Russell, Bertrand. *A Critical Exposition of the Philosophy of Leibniz: With an Appendix of Leading Passages*. Nottingham: Spokesman, 2008.
Rutherford, Donald. *Leibniz and the Rational Order of Nature*, Cambridge: Cambridge University Press, 2003.

Sabra, Abdelhamid Ibrahim. *Theories of Light: From Descartes to Newton*. 1st ed. Cambridge: Cambridge University Press, 1981.

Sayce, Richard. 'Maniérisme et périodisation, quelque réflexions général', in *Renaissance, Maniérisme, Baroque*, ed. Library Philosophique J.Vrin. Paris: Library Philosophique J.Vrin, 1972.

Schaub, Mirjam. 'Eine Begehung von Leibnizens Kristallpalast am Ende der Theodizee', in *Hide and Seek. Das Spiel von Transparenz und Opazität*, ed. Markus Rautzenberg and Andreas Wollsteiner. Munich: Fink 2010, 139–155.

Schaub, Mirjam. *Gilles Deleuze im Kino: das Sichtbare und das Sagbare*. 2. Aufl. München: Fink, 2006.

Schaub, Mirjam. *Gilles Deleuze im Wunderland: Zeit- als Ereignisphilosophie*. München: Fink, 2003.

Schaub, Mirjam. 'Mirrors as Elementar Optical Machines in the Work of Olafur Eliasson | Barocker Purismus. Spiegel als elementare Maschinen im Werk von Olafur Eliasson. Ausstellungskatalog zu "Baroque, Baroque" vom 221.11. 2015–06.03.2016 im Winterpalais des Prinzen Eugen in Wien', hg. von Daniela Zyman/T-B A21 Sternberg Press, 2015, 66–103.

Schepers, Heinrich. 'Monadic Perception', in *The Oxford Handbook of Leibniz*, ed. Maria Rosa Antognazza. Oxford: Oxford University Press, 2018, 381–92.

Schepers, Heinrich. 'Space and Time', in *The Oxford Handbook of Leibniz*, ed. Maria R. Antognazza. Oxford: Oxford University Press, 2018, 410–24.

Scherer, Jacques. *Le "Livre" de Mallarmé*. Paris: Gallimard, 1998.

Schmitt, Carl. *Political Theology: Four Chapters on the Concept of Sovereignty*. Chicago: University of Chicago Press, 2005.

Scholem, Gershom, and Lee Siegel. *Walter Benjamin: The Story of a Friendship*. New York Review Books Classics. New York: New York Review Books, 2001.

Schöttker, Detlev. 'Reduktion und Montage', in *Global Benjamin: Internationaler Walter-Benjamin-Kongress 1992*, ed. Wilhelm Fink Verlag. München: Wilhelm Fink Verlag, 1999, bd.2, 772–3.

Schröder, Peter. 'Carl Schmitt's Appropreation of the Early Modern European Tradition of Political Thought on the State and Interstate Relations', *History of Political Thought* 33, no. 2 (2012): 359.

Schwebel, Paula. 'Constellation and Expression in Benjamin and Leibniz', in *Thinking in Constellations: Walter Benjamin and the Humanities*, ed. Caroline Sauter and Nassima Saraoui. Cambridge: Cambridge Scholars Publishing, 2017, 51–82.

Schwebel, Paula. 'Monad and Time, Reading Leibniz with Benjamin and Heidegger', in *Sparks Will Fly: Benjamin and Heidegger*, ed. Andrew Benjamin and Dimitris Vardoulakis. Albany: State University of New York Press, 2015, 123–44.

Schwebel, Paula. *Walter Benjamin's Monadology*, Doctoral Thesis, Department of Philosophy, University of Toronto, 2012.

Sellars, John. 'Gilles Deleuze and the History of Philosophy', *British Journal for the History of Philosophy* 15, no. 3 (2007): 556.

Serres, Michel. *Le système de Leibniz et ses modèles mathématiques: Etoiles, schémas, points*. 3rd éd. Épiméthée. Paris: Presses Universitaires de France, 1990.
Sibertin-blanc, Guillaume. 'Pli psychotique et maniérisme, sur le leibnizianisme clinique de gilles deleuze', in *Gilles Deleuze, la logique du sensible – Esthétique & clinique*, ed. Adnen Jdey. Paris: Editions de l'incidence, 2013, 357–74.
Simmons, Allison. 'Changing the Cartesian Mind: Leibniz on Sensation, Representation and Consciousness', *The Philosophical Review* 110, no. 1 (January 2001): 31–75.
Smith, Daniel W. 'Deleuze on Leibniz: Difference, Continuity, and the Calculus', in *Essays on Deleuze*. Edinburgh: Edinburgh University Press, 2012, 43–58.
Smith, Daniel W. 'Genesis and Difference: Deleuze, Maïmon, and the Post-Kantian Reading of Leibniz', in *Deleuze and the Fold: A Critical Reader*, ed. Sjoerd van Tuinen and Niamh McDonnell. Basingstoke: Palgrave Macmillan, 2009, 132–54.
Smith, Daniel W., and Gilles Deleuze. *Essays on Deleuze*. Edinburgh: Edinburgh University Press, 2012.
Smith, Justin E. H. *Divine Machines: Leibniz and the Sciences of Life*. Princeton, NJ: Princeton University Press, 2011.
Smith, Justin E. H. *Irrationality a History of the Darkside of Reason*. Princeton, NJ: Princeton University Press 2019.
Snyder, Jon R. 'Art and Truth in Baroque Italy, or the Case of Emanuele Tesauro's Il cannocchiale aristoteli', *MLN* 131, no. 1 (2016): 74–96.
Spinoza, Benedictus de. *On the Improvement of the Understanding*, translated by Robert Harvey Monro Elwes, Dover Books on Western Philosophy. New York: Dover Publ, 1970, 372–3.
Spinoza, Benedictus de, and Stuart Hampshire. *Ethics*, edited by Edwin M. Curley. Penguin Classics. London: Penguin, 1996.
Spitz, Lewis W. 'The Significance of Leibniz for Historiography', *Journal of the History of Ideas* 13, no. 3 (June 1952): 333–48.
Steyerl, Hito. 'In Free Fall: A Thought Experiment on Vertical Perspective', in *The Wretched of the Screen*, ed. Julieta Aranda, Anton Vidokle and Brian Kuan Wood. Berlin: Sternberg Press, 2012, 12–30.
Steyerl, Hito. 'Mean Images', NLR, no. 140/141, March/June 2023.
Strong, Tracy B. 'Forward', in *Political Theology: Four Chapters on the Concept of Sovereignty* Chicago: University of Chicago Press, 2005.
Swoyer, Chris. 'Leibnizian Expression', *Journal of the History of Philosophy* 33 (1995): 65–99.
Szendi, Peter. *Le supermarché du visible*. Paris: Minuit, 2017.
Tagliacozzo, Tamara. *Experience and Infinite Task: Knowledge, Language, and Messianism in the Philosophy of Walter Benjamin*. Founding Critical Theory. London; Lanham: Rowman & Littlefield International, 2018.
Tissandier, Alex. *Affirming Divergence: Deleuze's Reading of Leibniz*. Edinburgh: Edinburgh University Press, 2022.
Tretyakov, Sergei. 'The Writer and the Socialist Village', *October* 118 (Fall 2006): 63–70.
Tuinen, Sjoerd van. *Mannerism in Philosophy*, Doctoral Thesis, Faculty of Arts and Philosophy, University of Ghent (2009).

Tuinen, Sjoerd van, and Niamh McDonnell, eds. *Deleuze and the Fold: A Critical Reader*. Basingstoke; New York: Palgrave Macmillan, 2010.

Valéry, Paul. *Variété*. Paris: Gallimard, 1998.

Valla Laurentius. 'Dialogue on Free Will', translated by Charles Edward Trinkaus in *The Renaissance Philosophy of Man*, Ernst Cassirer et al. (eds.). Chicago: University of Chicago Press, 1948.

Villani, Arnaud. 'The Problem of an Immanent Metaphysics', in *Gilles Deleuze and Metaphysics*, ed. Alain Beaulieu. Lanham, MD: Lexington Books, 2014, vii–x.

Voight, Frank et al. eds. *Material Und Begriff: Arbeitsverfahren Und Theoretische Beziehungen Walter Benjamins*. Deutsche Originalausgabe, Erste Auflage. Argument Sonderband. Neue Folge, AS 322. Hamburg: Argument, 2019.

Voltaire. *Candide*. London: Penguin, 2001.

Voss, Daniela. 'Maimon and Deleuze: The Viewpoint of Internal Genesis and the Concept of Differentials', *Parhesia* 11 (2011): 63.

Weber, Samuel. *Benjamin's -Abilities*. Cambridge, MA; London: Harvard University Press, 2010.

Weber, Samuel. 'Genealogy of Modernity: History, Myth and Allegory in Benjamin's Origin of the German Mourning Play', *MLN* 106, no. 3, German Issue (April 1991): 465–500.

Weber, Samuel. 'Taking Exception to Decision: Walter Benjamin and Carl Schmitt', *Diacritics* 22, no. 3/4, Commemorating Walter Benjamin (Autumn–Winter, 1992): 5–18.

Weigel, Sigrid. *Body-and Image-Space: Re-Reading Walter Benjamin*. Warwick Studies in European Philosophy. London; New York: Routledge, 1996.

Weizman, Eyal. *Hollow Land: Israel's Architecture of Occupation*. New edition. London New York: Verso, 2017.

Wohlfarth, Irving. 'Et Cetera? The Historian as Chiffonnier', *New German Critique* 39, Second Special Issue on Walter Benjamin (Autumn, 1986): 142–68.

Wohlfarth, Irving. 'Walter Benjamin's Image of Interpretation', *New German Critique* 17, Special Walter Benjamin Issue (Spring, 1979): 70–98.

Wölfflin, Heinrich. *Renaissance und Barock: Eine Untersuchung über Wesen und Entstehung des Barockstils in Italien*. Bruckmann: München, 1888.

Wong, Jack. 'Remapping the Constellation of Walter Benjamin's Allegorical Method', *American, British and Canadian Studies* 25, no. 1 (2016): 37–59.

Zalasiewicz, Jan, Colin N. Waters, Mark Williams and Colin P. Summerhayes, eds. *The Anthropocene as a Geological Time Unit: A Guide to the Scientific Evidence and Current Debate*. Cambridge; New York; Port Melbourne; New Delhi; Singapore: Cambridge University Press, 2019.

Zimmermann, Yvonne. 'Videoconferencing and the Uncanny Encounter with Oneself: Self-Reflexivity as Self-Monitoring 2.0', in *Pandemic Media: Preliminary Notes toward an Inventory*, ed. Philipp Dominik Keidl et al., Configurations of Film. Lüneburg: meson press, 2020, 100–3.

Acknowledgements

This book is based on research conducted for my doctoral dissertation at the *Centre for Research in Modern European Philosophy* (*CRMEP*) at Kingston University, London. I am extremely grateful for the encouragement, patience, thoughtful comments and discerning critique of my first supervisor, Peter Osborne. I am thankful for his introducing me to the philosophical métier and teaching me most of what I know about philosophical argumentation and critically approaching the history of philosophy. I wish to warmly thank my second supervisor, Éric Alliez, for helping me believe in my ability to venture into uncharted territory, and for his valuable comments and suggestions. I also would like to thank Lucie Mercier, who accompanied the project over the four years of its development with perceptive comments. In my first year at the (*CRMEP*) I had the pleasure of being a teaching assistant in two memorable courses: Étienne Balibar's course on Spinoza's *Ethics* and Stella Sandford's course on Kant's First Critique. I wish to thank them, as well as Andrew Benjamin, Peter Hallward and Catherine Malabou, for their continuous support and intellectually stimulating lectures and classes. I would also like to warmly thank my thesis examiners, Howard Caygill and Mirjam Schaub, for their thoughtful comments on my dissertation and for encouraging me to turn it into a book.

My fourth year of writing was spent on an exchange in Berlin. I wish to thank my generous and welcoming hosts in the city; Daniel Weidner from the *Leibniz-Zentrum für Literatur- und Kulturforschung (ZfL)*, Anne Eusterschulte from the *Institut für Philosophie* at the *Freie Universität* and Katia Genel from the *Centre Marc Bloch* (*CMB*). They have all taken time to read and reflect on my work and I have benefited from the multiplicity of their perspectives. I am grateful, too, for the help and guidance of Jakob Vogel and Aurélie Denoyer from the *CMB*. Ursula Marx from the *Walter Benjamin Archive* has made me feel at home at the archive from my first visit.

Many of my ideas were developed through conversations and debates with friends and colleagues. I would like to thank Tom Vandeputte, Nassima Sahraoui and Stephano Marchesoni for organizing two wonderful workshops, in which I had the pleasure of taking part, on the concepts of 'perception' and 'intensity' in Walter Benjamin's writing. I also would like to thank Chris Law, with whom I co-organized the workshop 'On Expression: Benjamin and Leibniz' at Goldsmith University, London, and Frank Müller, co-organizer of the workshop 'Übersetzungen von Gewalt. Walter Benjamins "Zur Kritik der Gewalt"' at *CMB*. Both of these were stimulating and inspiring occasions for the exchange of ideas from which my own work profited. I would also like to thank Alexandra Richter, Philip Eckhardt and Ilit Ferber for their guidance and feedback; and Eli Friedlander, who first introduced me to Walter Benjamin during my BA studies.

My gratitude goes to Orna, Gidi and Dan Hadari and to Amira and Alexander Goehr, who hosted me generously in their beautiful homes during intensive writing periods. I am deeply thankful to friends in Berlin, Tel Aviv and London: in particular Peter Taoussanis, Daniel Nemenyi, Daniel Weizman, John Douglas Millar, Marie Louise Krogh, Isabell Dahms, Tamara Micner, Josh Weiner, Dafna Sharabi, Alex Cocotas, Matan Nahaloni, Tamar Kay and Zehavit Zaslanski. I would also like to thank my family; special thanks go to my father Gabriel, who bought me a copy of Benjamin's *Arcades Project* for my sixteenth birthday and read many a draft over the years, my mother Anat, my brother Boaz and to Kevin for their loving encouragement.

**

This book is based on my doctoral dissertation which was generously supported by a Kingston University scholarship between 2015 and 2018, a scholarship from The *Studienstiftung des Abgeordnetenhauses von Berlin* during 2018–19 and a grant from the *Centre Marc Bloch* between October and December 2019.

Index

www.ingramcontent.com/pod-product-compliance
Ingram Content Group UK Ltd.
Pitfield, Milton Keynes, MK11 3LW, UK
UKHW020804290926
15470UKWH00003B/950

9 781350 414211